socioeconomic inequalities and health

Proceedings of the
Socioeconomic Inequalities and Health Conference,
Wellington, 9-10 December, 1996

Edited by
Peter Crampton and Philippa Howden-Chapman

First published in 1997

Institute of Policy Studies
Victoria University of Wellington
PO Box 600
Wellington
New Zealand

© The Institute of Policy Studies

ISBN 0-908935-21-8

Cover design by Jacob Sullivan
Printed by The Printing Press,
Thorndon Quay, Wellington

Contents

1 ~ Introduction

Philippa Howden-Chapman and Peter Crampton

The research links between socioeconomic status (SES) and health are well-established. There is continuing debate around causality and the added contribution of ethnicity and gender, but few now seriously challenge the fundamental importance of socioeconomic factors for health status.

This is a key area for public health, yet there is no public health area which is more controversial. Research on socioeconomic status and health presents a dilemma for researchers, policy-makers and politicians. For researchers the issues have intrinsic interest, but as the evidence produced both here and overseas seems impotent to affect public policy significantly, the challenge is to strengthen the evidence further and make the policy implications clearer. For politicians, both those in government and those outside, a major challenge is presented by the increasingly clear evidence of the impact of income disparities both at an individual and community level.

This conference was planned with two aims:

i. to facilitate the linkage between research and policy; and
ii. to give researchers the opportunity to meet and consider collaborations in time for the up-coming Health Research Council call for research proposals in this area.

The exchange of ideas and the building of trust for a common purpose and the public good were critical ingredients of the conference, and the broad range of participants from the community, universities, government and Maori, and the generosity and frankness of presenters, commentators, chairpersons and participants were a measure of the success of the conference.

One of the key themes was the rigorous scrutiny of the evidence. The keynote overseas speakers, Peter Saunders and Ichiro Kawachi, examined the cross-sectional evidence on the impact of income inequalities on health. The debate focused on the impact of the size of the income gap between rich and poor on health.

2 ~ Socioeconomic Inequalities and Health

The principal challenge for the conference was to place research in the context of social policy. Both Peter Saunders and commentator Brian Easton emphasised the critical role of high quality social science research in the formulation of policies which properly address the health needs of populations. The example mentioned above, that of the impact of widening income differentials on health, is of major importance considering the emphasis placed on constraining benefit entitlements, increasingly less progressive taxes and a reduced social wage in New Zealand's economic and social policy over recent times. George Barker outlined evidence that emphasised individuals' lifetime earnings and offered a complementary picture of social policy development from the point of view of a Treasury manager.

Commentators Susan St John, Cindy Kiro and Alison Blaiklock responded by placing research in the context of political philosophy and outlined in different ways the seriousness of public health and social problems that need to be addressed by policy-makers.

It is imperative that public health and social policy researchers translate their findings into useable tools and theories. Three groups of interdisciplinary researchers gave papers on such measures, focusing in turn at the level of a single population, a small geographical area and an individual measure of occupation. Bob Stephens and Charles Waldegrave presented their poverty line research, which offers a benchmark which may be used to assess the effects of social policy changes on the poor. Peter Crampton, Claire Salmond and Frances Sutton similarly offered a new area-based index of deprivation – *NZDep91* – which is intended to assist in policy formulation and the commissioning of health services. Philippa Howden-Chapman, Keith McLeod and Peter Davis presented research on a new occupational index of socioeconomic status which up-dates the Elley-Irving Scale.

This book contains all the papers from the conference in the order in which they were presented. Papers from the commentators have been included, providing critical and interpretative insights. Edited summaries of the three workshop streams are included at the end of the book.

These summaries, which reflect comments from the wide range of conference participants, provide useful suggestions for researchers and policy-makers for addressing many of the issues raised in the conference. To facilitate future networking, a full list of participants is appended. The conference was generously sponsored by a range of organisations.

This reflects the broad commitment to improving public health in New Zealand. The sponsors were: the Public Health Association, the Health Services Research Centre (funded by the Health Research Council and affiliated with both Victoria University and the Wellington School of Medicine), the Department of Public Health at the Wellington School of Medicine, and the National Advisory Committee on Health and Disability Services.

2 ~ Setting the Scene

Alistair Woodward

Compare these people, bent over their looms each day, arising in shadows, wilting, one might say, like plants; compare them to other residents of the same places or to farmers who work and live in the open air with bright sunlight, and you will be amazed by the difference. This difference is enormous and is well known to the officials in charge of army recruitment . ... In order to find 100 men fit for military service, 193 men have to be drafted in the more comfortable classes and up to 343 have to be drafted in the poorer classes. (Villerme, 1840)

This is familiar territory to those who have studied public health. The severe contrasts in wealth and health that resulted from or, some would say, were brought to light by the Industrial Revolution in Europe in the early 19th century gave cause and reason to the renaissance of public health. Generations of budding epidemiologists have learnt their craft by studying comparisons of the kind reported by Villerme.

However, when considering social inequalities in contemporary New Zealand it would be unwise to take anything for granted. There are many views, firmly held and widely divergent. This has been made quite clear by recent arguments about the levels of poverty in New Zealand. One study, described in Robert Stephens' and Charles Waldegrave's paper in these proceedings, has claimed that 21% of the population falls below the poverty line. Jenny Shipley, Acting Social Welfare Minister at the time the study was released, sees the world differently – "to claim that New Zealanders are living in poverty in 1996, given all the changes the Government has made, I think is highly misleading" (*Evening Post*, 12 April 1996).

It may be helpful, when trying to connect research to policy, to attempt answers to these questions: are social inequalities real? do they matter? are they avoidable? and, if avoidable, where to begin?

The first task should not be too challenging as data are plentiful, but whether data are accepted is another matter. The rationale for tackling social inequalities and health deserves careful argument. As Peter Saunders

5

pointed out to the conference, it is not enough to rely on egalitarian instincts; we have to present a very good case for why inequalities matter. It may be important to be clear that "reducing inequalities in health" does not mean levelling down, but levelling up. "Inequalities" in the research literature tends to be used cryptically, as a kind of shorthand for factors that cause excessive and avoidable disease and injury.

It is important to emphasise that the burden of illhealth associated with social disadvantage is onerous. For example, if all New Zealand males aged 15-64 years experienced the same death rates as those of the most privileged in the country, the mortality rate in this group would fall by 40% (see Pearce, Marshall and Borman, 1991). There is some evidence that the effects of disadvantage on health are felt by all, not only those at the bottom of the social pile. It is not entirely clear why the steepness of the social class gradient should affect the whole population, but this finding has been reported from several countries, including work carried out in the United States by Ichiro Kawachi and his colleagues.

Inequality is essentially a social phenomenon and it is important to bear this in mind, although past experience and research culture incline many of us to thinking predominantly of causes of illness that operate at the level of individuals. As an antidote, I recommend Richard Leakey's most recent book (1996), *The Sixth Extinction*. Leakey points out that a failure to think at a social or ecological level has led to severe misundertanding of the reasons why some species prosper and others fail.

In the past it was thought that the success of species depended solely on their competitive advantage over other species occupying comparable ecological niches. In fact this view does not fit well with what is observed to occur – would-be invaders may fail to take hold in a host community, even where the invader is superior to the species it attempts to displace. The explanation appears to be that 'competitive advantage' is not only a property of species, it is also a property of the communities in which species are located. Apparently inferior species, embedded in mature communities, are protected by networks and interactions that operate at a higher level of complexity. There is nothing mystical or ideological about this – the connections between species, for example, through food chains, provide very practical mechanisms by which the whole is made more than the sum of the parts.

Perhaps this notion of 'advantage' has useful applications in studies of the health of human populations. For example, it seems a positive step to move beyond analyses of social roles (as measured at the level of individuals) to the notion of social capital (a characteristic of groups or populations that encourages people to cooperate for their mutual benefit).

It seems a pity, as Ichiro Kawachi has observed, that we know more about how to lose this precious commodity than how to gain it. I trust this conference and the proceedings that result from it will be a step towards re-building social capital, by contributing to a better understanding and wider appreciation of the influence that social inequalities has on the health of New Zealanders.

References

Evening Post (Wellington), 12 April 1996, p 2

Leakey, R and R Lewin (1996) *The Sixth Extinction: Biodiversity and its Survival*, London: Weidenfeld and Nicolson

Pearce, NE, S Marshall and B Borman (1991) 'Undiminished social class mortality differences in New Zealand men', *New Zealand Medical Journal*, 104, pp 153-156

Villerme, LR (1840) *Tableau de L'Etat Physique et Moral des Ouvriers Employes dans les manufactures de Coton, de Laine et de Soie*, Paris: Jules Renouard et Cie, Libraires, excerpted and translated in C Buck, A Llopis, E Najera and M Terris (eds) *The Challenge of Epidemiology. Issues and Selected Readings*, Washington: World Health Organisation, 1988, pp 33-37

3 ~ Do Inequalities in Income Cause Inequalities in Health?

Peter Saunders

Introduction

It is well known that material living standards exert an influence on health status, even though the effect may be weak when comparing experience across high income countries or the health of those with high incomes within individual countries. The policy relevance, and even the existence, of this relationship has, however, been challenged in a series of recent papers which have argued that *relative* income and circumstances may exert a stronger impact on health than the *absolute* level of income or well-being. From this, it follows that the extent of inequality in health status will depend upon the degree of income inequality among that population, whatever the level of its mean income.

One of the leading proponents of this line of research, Richard Wilkinson, concludes his much-cited study of the relationship between income distribution and mortality by claiming that the evidence:

> ... suggests that the association between health and income distribution is a result of factors to do with relative rather than absolute income. ... A shift in emphasis from absolute to relative standards indicates a fall in the importance of the direct physical effects of material circumstances relative to psycho-social influences. The social consequences of people's differing circumstances in terms of stress, self esteem, and social relations may now be one of the most important influences on health. (Wilkinson, 1992, p 169)

If these claims are correct, they imply that increases in income inequality will have adverse consequences for disparities in health and, to the extent that these lead to increased pressure on health resources, will lead in turn to increases in health spending. The result may give rise to a contradiction between the trend to economic liberalism and the policies of strict fiscal restraint which form part of it.

Furthermore, the latest international evidence on trends in income inequality in OECD countries (see Atkinson, Rainwater and Smeeding, 1995) and former state socialist countries of Central and Eastern Europe (see Torrey, Smeeding and Bailey, 1996) has highlighted a tendency for income inequality to be rising in most – though not all – of them. Smeeding (1996) summarises the evidence in the following terms:

> Our overall assessment is that any 1970s trend towards greater *equality* has ended in virtually all of the nations studied here, with the single exception of Italy. And there is a tendency for those nations with the most recent data to show rising disposable income inequality. Certainly, market income trends show this type of change in most nations observed here. It may well be that the nations that have so far shown resistance to rising disposable income inequality may soon exhibit such trends. (Smeeding, 1996, p 51; italics in the original)

To the extent that most nations have experienced a 'rising tide of inequality' in the 1980s and 1990s, the implications for social justice as well as for health will present a fundamental challenge for the years ahead.

Given their public policy significance, it is not surprising that the empirical findings of Wilkinson have been subject to critical study by other researchers (e.g. Judge, 1995; Saunders, 1996). However, the issues raised by these authors appear not to have affected the view of those who have argued that the existence of the empirical associations highlighted in the work of Wilkinson and others (e.g. Rodgers, 1979; Le Grand, 1987) "seems secure" (Smith, 1996).

Meanwhile, two recent American studies of state differences appear to support the existence of an empirical association between the degree of income inequality and a range of mortality measures (Kaplan et al, 1996) as well as between inequality and several other indicators of social distress (Kennedy, Kawachi and Prothrow-Stith, 1996).

This paper addresses two aspects of the debate over the relationship between income distribution and inequalities in health. The first, addressed in the next two sections, focuses on Wilkinson's own research and the literature which it has generated. It challenges the claim that the statistical associations revealed in this literature have established causality between income inequality and health, mainly because the studies on which they are based provide no guide to the causal pathways of cause and effect.

This is followed by an overview of some of the results from a recent study (Saunders, 1996) which explores the impact of relative economic status on health using data on a large cross-section of Australian households. The next section addresses some of the broader social policy implications of this body of research, followed by a summary.

Income Distribution and Mortality: Revisiting Wilkinson

A common theme to emerge from recent cross-national research on income distribution and mortality is that there exists a reasonably clear and stable cross-national relationship between these two variables. Thus, McIsaac and Wilkinson have recently concluded that:

> ... at least eight different research workers or groups have reported statistically significant relationships between income distribution and measures of mortality. (McIsaac and Wilkinson, 1995, p 3)

Although 'many correlations do not a causation make', the fact that many studies have uncovered the same statistical association suggests that causal mechanisms may exist, even if they remain for the moment undiscovered.

The findings themselves have already had a major impact on a number of national and international agencies concerned with issues associated with health. Thus, for example, the World Bank asserts in its 1993 *World Development Report* that:

> The higher a country's average income per capita, the more likely its people are to live longer and healthier lives. Of course, this effect tapers off as income rises ... Because poverty has a powerful influence on health, it is not just income per capita that is relevant; the distribution of income and the number of people in poverty matter as well. *In industrial countries life expectancy depends much more on income distribution than on income per capita, and it has been rising faster in countries with improving income distribution.* (World Bank, 1993, pp 39-40; italics added)

For an organisation whose conservative economic analysis and policy prescriptions have frequently come under attack, it is surprising that the same rigour has not been applied to such assertions, the implications of which are both radical and far-reaching.

Much of the recent cross-country evidence establishing the link between income distribution and life expectancy is based on Wilkinson's 1992 article in the *British Medical Journal* (Wilkinson, 1992). The results from this study have been widely cited by others and have been emphasised in a series of subsequent papers published by Wilkinson (e.g. 1993, 1994).

In his original paper, Wilkinson begins by noting the existence of a well-established relationship between the level of income in a country and several health indicators including life expectancy at birth, and then explores for a sub-set of OECD countries whether there is any relationship between life expectancy and alternative measures of the degree of inequality of family income. This analysis was restricted to the nine countries for which distributional data were available from the Luxembourg Income Study (LIS) – Australia, Canada, Germany, the Netherlands, Norway, Sweden, Switzerland, the United Kingdom and the United States – and covered years around 1980, the 'first round' of the time when the LIS dataset was assembled.

Wilkinson's analysis appeared to show that life expectancy was correlated with two income inequality measures: the share of income received by each decile of the distribution, and the cumulative income share accruing to each successive grouping of the deciles. The correlation coefficients reported by Wilkinson are shown in the first two columns of Table 3.1. On the basis of these results, Wilkinson concluded that the relationship was strongest when inequality was measured by the share of disposable income received by the lowest 70% of families. The relationship between these two variables, shown in Figure 3.1, provides a vivid illustration of the apparent strength of the identified relationship.

Having established the existence of the statistical association, Wilkinson argued that it did not reflect the impact of intervening variables, nor was there evidence to sustain the view that causality goes from mortality to inequality rather than the reverse. He concluded that, for the richer countries belonging to the OECD, the association between income distribution and health is a causal relationship and that it is relative rather than absolute income which matters for health. His analysis concludes with the following remarks:

Table 3.1: Life Expectancy and Income Distribution: Re-Working Wilkinson

(Pearson correlation coefficients)

	Wilkinson(b) Table 1		Correlation coefficients:(a) Replicated(b) results		Revised(b) results	
Disposable income share	I	II	I	II	I	II
Lowest 10%	0.13	0.13	0.16	0.16	-0.02	-0.02
Lowest 20%	0.63*	0.45	0.68**	0.49*	0.48	0.28
Lowest 30%	0.76**	0.57*	0.81***	0.61*	0.44	0.34
Lowest 40%	0.92***	0.65**	0.95***	0.70*	0.30	0.34
Lowest 50%	0.64**	0.75**	0.61*	0.79***	-0.03	0.31
Lowest 60%	0.28	0.84***	0.23	0.87***	-0.29	0.24
Lowest 70%	0.18	0.86***	0.12	0.87***	-0.41	0.13
Lowest 80%	0.17	0.80**	0.12	0.79**	-0.44	0.09
Lowest 90%	0.11	0.68*	0.03	0.65*	-0.48	-0.11

Note: (a) The asterisks (*/**/***) indicate statistical significance at the (0.05/0.01/0.001) level.
(b) The correlation coefficients shown in the columns marked as I refer to those between each separate decile share and life expectancy. Those in the columns marked as II refer to the correlations between the cumulative decile shares and life expectancy.

Sources: Wilkinson, 1992, Table 1; Bishop, Formby and Smith, 1991, Table 2.

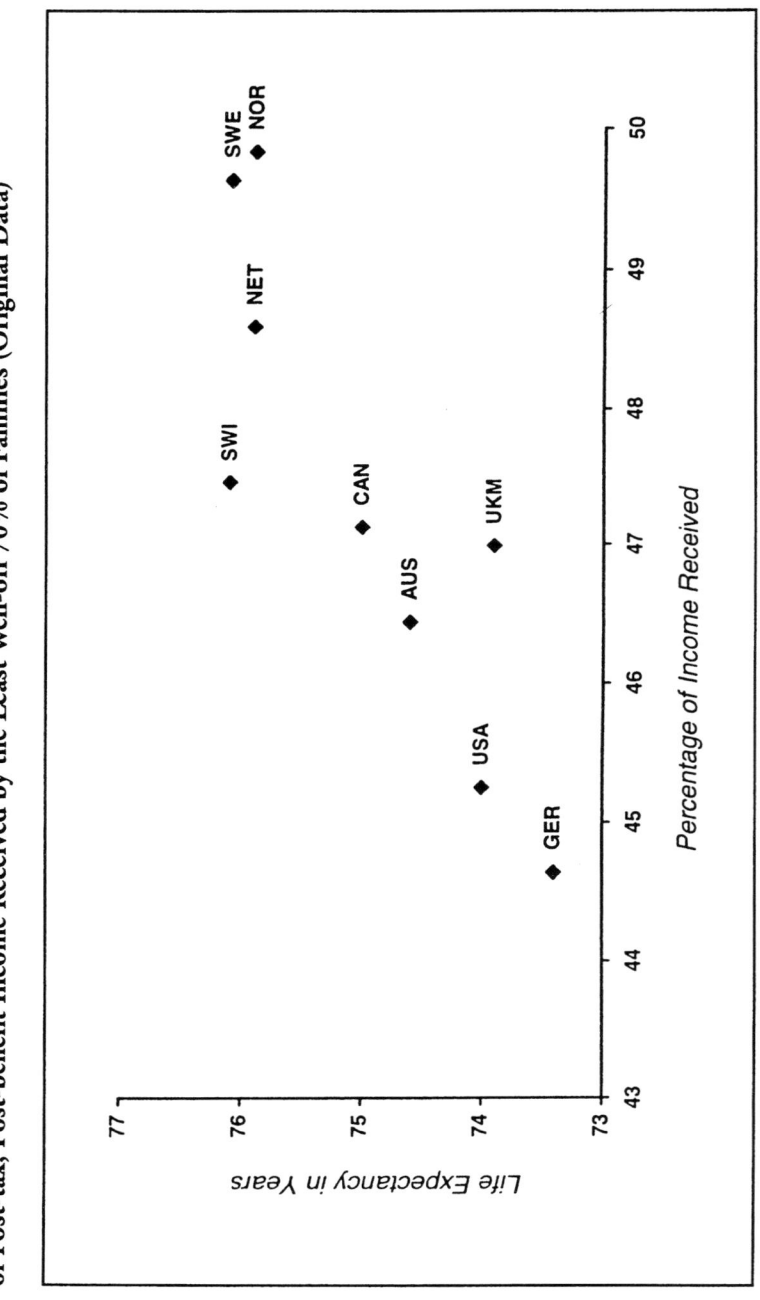

Figure 3.1: The Relation Between Life Expectancy at Birth (Male and Female Combined) and the Percentage of Post-tax, Post-benefit Income Received by the Least Well-off 70% of Families (Original Data)

... the sense of relative deprivation, at being at a disadvantage relative to those better off, probably extends beyond the conventional boundaries of poverty ... the social consequences of people's differing circumstances in terms of stress, self-esteem and social relations may now be one of the most important influences on health. (Wilkinson, 1992, p 168)

It is worth noting at this stage that the inequality measure shown by Wilkinson to be most highly correlated with life expectancy – the income share of the lowest 70% of families – is somewhat unusual. It is not one included among conventional inequality measures (as outlined in Cowell, 1995, for example), nor does it conform to many of the axioms regarded as desirable properties of relative inequality measures (Jenkins, 1991). Nor, more significantly, is any attempt made to relate the choice of measure to the processes through which its impacts are presumed to occur.

In light of these considerations, it is interesting to explore whether the main features of Wilkinson's results are sensitive to the inequality measure used and, if so, how. In exploring this issue, the first point to note is that the distributional data used by Wilkinson were substantially revised by those associated with the LIS project. The precise nature of these adjustments is explained by the authors of the LIS study on which Wilkinson had relied when their results were eventually published two years after they first appeared in Working Paper form (Bishop, Formby and Smith, 1991).

The first Note to the published paper reads as follows:

An earlier version of this paper was widely circulated as LIS-CEPS Working Paper no. 26 (1989). Since the empirical work for the Working Paper was completed, several important modifications have been made to the country-specific LIS data-sets used in this study. First a number of outlying observations in the upper tail of the German data-set have been changed to make for greater comparability with other LIS data. Second, one US observation with a tax liability of over $300,000 has been dropped. Third, the number of Swiss zero and negative incomes has reduced. Finally, changes in the Netherlands and Australian data have also been made. The changes in the German and Swiss data appear to be particularly important. (Bishop, Formby and Smith, 1991, p 475)

The differences that arise if the revised LIS data are substituted for those originally used by Wilkinson are shown in the final two columns of Table 3.1 (the middle two columns confirm that Wilkinson's original results can be closely replicated). The difference this makes to the pattern of results is remarkable. There is now no evidence of a statistically significant association between the decile shares of income and life expectancy, nor between the cumulative income shares and life expectancy.

The difference these data modifications make to the results is illustrated in Figure 3.2. The effect of incorporating the data revisions is to reduce greatly the degree of measured inequality in Germany and to increase it greatly in Switzerland. These changes cause the cross-sectional relationship between income distribution and life expectancy virtually to disappear. There is no longer any evidence of a statistically significant correlation between the two variables, as the actual correlation coefficients reported in the final two columns of Table 3.1 confirm.

Other writers (e.g. Forster, 1992) have previously cast doubt over the validity of Wilkinson's results. Most notably, Judge (1995) provides a spirited critique of Wilkinson's data and methodology, arguing that:

> ... the strength of the relation between income inequality and average life expectancy has been exaggerated ... [and that] ... a careful review of the evidence does not support the hypothesis that inequalities in income distribution largely explain differences in average life expectancy among rich countries. In retrospect, it seems extraordinary that a predominantly monocausal explanation of international variations in life expectancy should ever have been regarded as plausible. (Judge, 1995, p 1284)

Three issues raised by Judge are worth emphasising. The first relates to the fact that Wilkinson did not adjust his data on family incomes for differences in family size – the equivalence scale issue. It is now universally acknowledged in the distributional literature that an equivalence adjustment is essential if estimates of the *distribution of income* are to be related to the *distribution of economic welfare*. A second point is that in

Figure 3.2: The Relation Between Life Expectancy at Birth (Male and Female Combined) and the Percentage of Post-tax, Post-benefit Income Received by the Least Well-off 70% of Families (Revised Data)

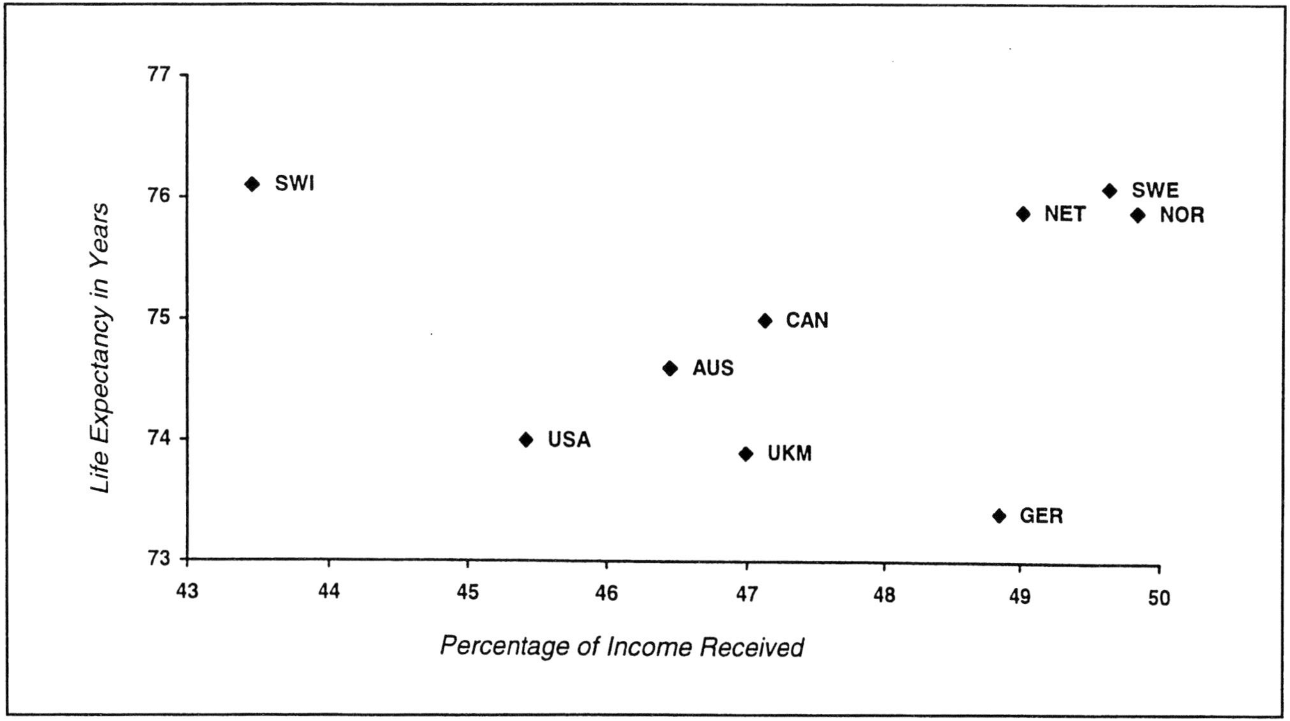

choosing which inequality measure to use, account should be taken of which measure might be expected to be associated with life expectancy and why. Finally, Judge demonstrates that the use of an equivalence adjusted income measure or the substitution of one inequality measure by another serves to render the association between income inequality and life expectancy no longer statistically significant.

Although Wilkinson acknowledged in his response to Judge (Wilkinson, 1995, p 1286) the problems associated with the 'revisions' to the LIS data, the main point he made was that the statistical association between income distribution and mortality had not only been observed by others in earlier work, but was also about to receive further confirmation in two forthcoming studies using cross-sectional data for the United States (see below). In drawing attention to these issues, Wilkinson never adequately addressed the criticisms of his data and methodology raised by Judge.

If Wilkinson is correct in asserting that the body of empirical evidence increasingly supports his position, then it ought to be possible to check this using new data. In attempting this, several different approaches were explored using the latest available distributional estimates derived from the LIS data. The source for these is the recent study of income distribution commissioned by the OECD (Atkinson, Rainwater and Smeeding, 1995). One advantage of using the results from this study is that they cover a broader range of OECD countries – including in some instances countries that are not included in the LIS dataset itself.

In general, the attempt to uncover a significant association between income inequality and life expectancy using the latest LIS distributional data proved unsuccessful. One particular set of results illustrates the general pattern of the findings which emerged. These were derived from a model in which inequality was measured by the ratio of the ninetieth (P90) to the tenth (P10) percentiles of the distribution of equivalent disposable family income among individuals, using an equivalence scale equal to the square root of total family size. Both the inequality measure and the equivalence scale have received a good deal of support for use in comparative research in the recent distributional literature.

Figures 3.3 and 3.4 show, respectively, the relationship between inequality and life expectancy in the mid-1980s (using the 'second wave' of LIS data) and changes in inequality and life expectancy over the first half of the 1980s (between the first and second LIS waves). In neither case is the relationship between the two variables significant. Nor is there one if

Figure 3.3: The Relation Between Life Expectancy at Birth (Male and Female Combined) and the Ratio of the Ninetieth to the Tenth Income Percentile in OECD Countries

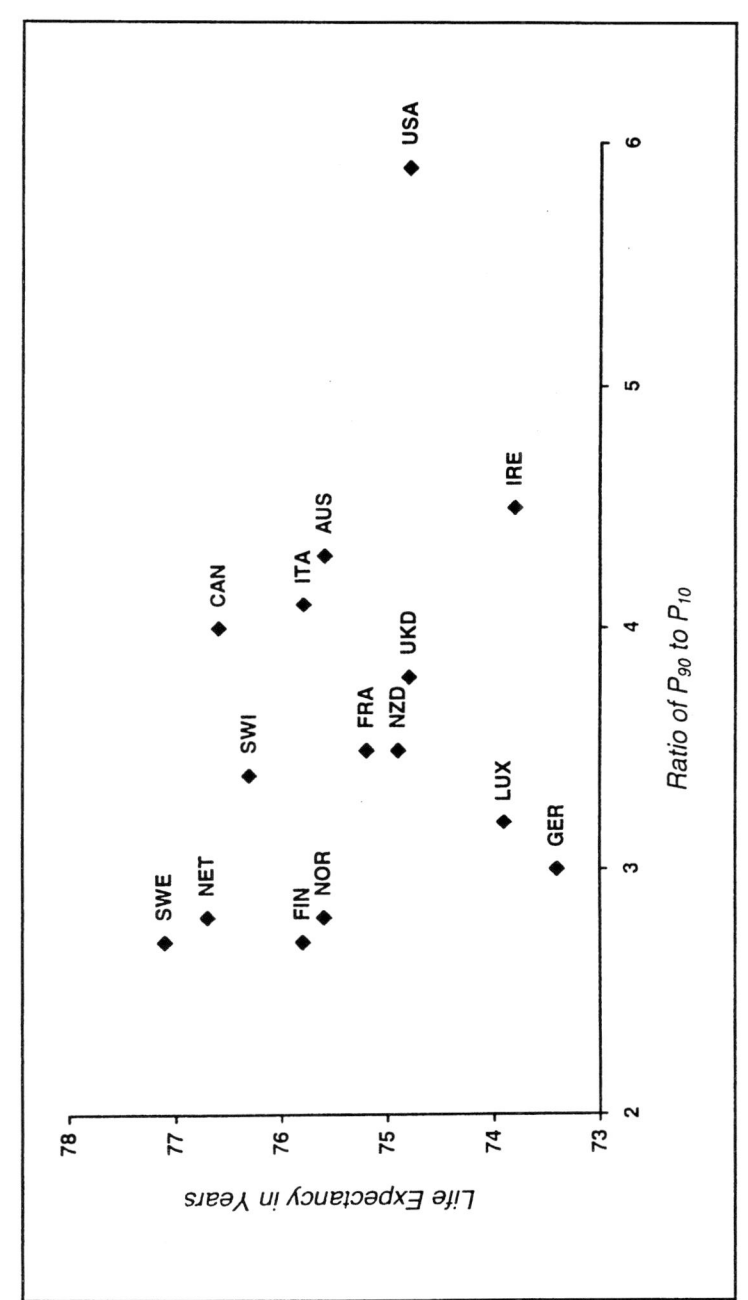

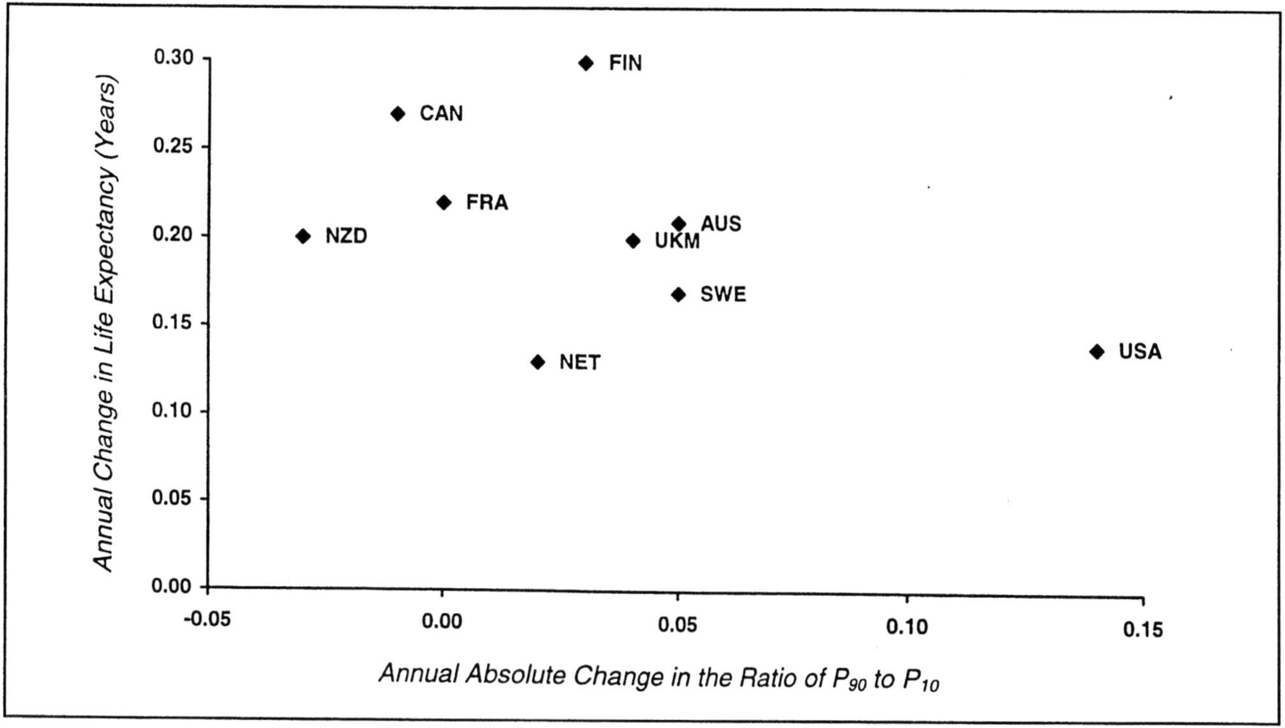

Figure 3.4: The Relation Between the Annual Rate of Change of Life Expectancy (in Years) and the Annual Absolute Change in the Ratio of the Ninetieth to the Tenth Income Percentile in OECD Countries

the Gini measure of inequality is used, or if a non-linear formulation is estimated. Nor does the relationship become significant after controlling for differences in the level of GDP and health expenditure (public, private or total) in each country (details are provided in Saunders, 1996).

These results thus cast doubt on the existence of a significant association between income distribution and life expectancy among OECD countries in the 1980s, when more sophisticated measures of inequality are used and after controlling for the impact of other relevant variables. Far from being 'secure', the association between income distribution and life expectancy, at least at the simple aggregative level explored here, is at best tenuous and certainly in need of further detailed study.

Reincarnation?

Two studies published earlier this year in the *British Medical Journal* (to which Wilkinson made reference in his reply to Judge) have breathed new life into the debate over inequality and health (Kennedy, Kawachi and Prothrow-Stith, 1996; Kaplan et al, 1996). Certainly, both studies take account of several of the data and methodological criticisms outlined above. In particular, the attempt to separate out the impact of income inequality on the different causes of death represents an important advance which Wilkinson's own more recent work also incorporates (McIsaac and Wilkinson, 1995).

Both studies use data from the 1990 United States census population and housing summary tape to explore the relationship between income inequality and mortality in different US states. Both also take care over the way in which they choose to measure the degree of inequality in the income distribution. Both studies conclude that state differences in total mortality and in specific causes of death are significantly related to differences in income inequality between the states.

However, both studies remain open to criticism. In particular, neither pays sufficient attention to which measures of inequality are best suited to explain differences in mortality in terms of an underlying causal model or theoretical structure which identifies and delineates chains of influence. Both use quite different measures, the Kaplan et al study using the income share of the least well-off 50% and the Kennedy et al study preferring the Robin Hood Index (RHI) which measures the share of total income which would have to be redistributed from those with incomes above the average to those with incomes below average in order to achieve complete equality.

Kennedy et al go as far as to acknowledge that the relatively arbitrary choice of inequality measure "creates the hazard that researchers will use the measure that proves the result they wish to find" (p 1007) but do not justify their use of the RHI except on the (unsubstantiated) claim that it "has a plausible interpretation" (p 1007), even though their own empirical results illustrate that the different measures of inequality within a *given* income distribution often have very low correlations with each other.

Aside from these measurement issues, both studies fail to address a number of the shortcomings that characterised Wilkinson's earlier work. Thus, both use the census income measure which makes no allowance for the payment of direct taxes, and neither makes an equivalence adjustment at the family level to account for differences in family size and hence need.

In this context, Kennedy et al's claim that –

> Although some researchers advocate the use of equivalence scales to take into account differences in household size, such scales ignore the effects of economies of scale (Kennedy et al, 1996, p 1006) –

can only be described as bizarre.

It is also notable that neither study is able to find evidence linking *changes* in inequality systematically to *changes* in mortality, even though Wilkinson (1992, p 166) saw this as providing "a more demanding test of the relation". In summary, while both studies address some of the limitations of the earlier work, neither can claim to have provided a solid empirical foundation to the relationship between income distribution and mortality. The best that can be claimed is that they produce more results which raise an intriguing set of questions concerning the social determinants of health, but do not succeed in answering them.

What needs to be done to advance this important debate is to give more detailed consideration to the processes and patterns of behaviour (what Kaplan et al refer to as the "potential pathways") which underlie the observed aggregate relationships. Kennedy et al acknowledge that income distribution may be acting as a proxy for other explanatory variables such as the degree of investment in human capital. In a similar vein, Wilkinson (1992, p 167) has identified the quality of public services as a potential intervening variable.

In addition to identifying and exploring the role of intervening variables, there is also a need to take account of the possibility of reverse causality in which low income results from poor health preventing or limiting labour market involvement rather than causing it. A good deal more thought also needs to be given to which health indicator conforms most closely with the underlying causal mechanisms and processes. Finally, some thought should be given to whether or not there is a need to adjust the inequality measures for differences in population age structure if, as both Kaplan et al and Kennedy et al do, an age-adjusted mortality measure is used.

These issues need to be given more attention if the sceptics are to be convinced that the statistical correlations are robust and can be related to what is causing them. This will involve undertaking studies which look directly into the underlying mechanisms to complement those which focus on aggregate relationships. Such an approach has consequences not only for the selection of an appropriate inequality measure, but also for the measure (or measures) of health status that are used. The next two sections present some results for Australia which adopt this approach to the issue.

Some Australian Results

This section summarises some of the results derived from a recent study of the relationship between income and health status in Australia (Saunders, 1996). The study utilises unit record data from the *National Health Survey* (NHS) conducted in 1989-90 which provides an enormous amount of detailed information on the health status of over 54,000 individuals living in 22,000 households. Although the NHS only provides gross annual income figures in ranges, unit record data from the *1990 Survey of Income and Housing Costs and Amenities* were used to impute a precise estimate of gross income from which disposable income was then derived.

These incomes estimates were then expressed relative to need by adjusting by an equivalence scale and used to explore how health status is related to variations in the level of equivalent income. This was done both by comparing the health status of those whose incomes placed them either side of a poverty threshold (described below) as well as by investigating how health status varies across the deciles of the distribution of equivalent disposable income. Only some examples of the former results are described here; those interested in the latter should consult Saunders (1996).

Before proceeding, it is important to emphasise that because the data are taken from a single cross-section, they provide no basis for assessing whether it is the absolute or relative level of income which exerts an influence on health. This is unfortunate given the importance that Wilkinson and others place upon the distinction between the two, but investigating this issue in detail requires the use of longitudinal data and the application of a methodological framework similar to that employed recently by Lundberg and Fritzell (1994).

To be fair to Wilkinson, it is also important to acknowledge that he is sceptical about using the poverty rate as a measure of inequality, partly because this can be sensitive to low response rate (or, more accurately, to under-reporting of income by those at the bottom of the distribution) and partly because he believes that the measure of inequality should not be confined to the bottom portion of the distribution (Wilkinson, 1995, p 1287).

On the other hand, Wilkinson himself makes reference to the role of relative poverty in leading to stress and a loss of self-esteem which eventually result in bad health (Wilkinson, 1992, p 168). Certainly, the literature on relative deprivation social exclusion suggests that poverty is more likely to be associated with these inhibiting characteristics than with broader inequality measures which encompass a larger section of the population.

Poverty status has been assessed using the poverty line framework proposed by the Commission of Inquiry into Poverty (1975) – the Henderson poverty line. Although this framework has been the subject of a good deal of recent criticism (e.g. from Harding and Mitchell, 1992; and Gruen, 1995), attention here focuses on the health status *differences* between those on either side of the poverty line, not on producing definitive estimates of the level of poverty or how it has changed over time. However, in acknowledgment of the criticism levelled at the use of the Henderson poverty line, the poverty threshold was set at a level 20% above the line itself – a measure which the Poverty Commission used to include both the poor and those on the margins of poverty – which also moves in the direction of a broader measure of relative inequality.

Poverty, Health and Happiness
By way of introduction to the NHS data, we begin by summarising responses to two questions relating to the subjective health status and overall well-being of the respondents. These two subjective evaluation questions were asked early in the survey, immediately following a series of

background questions on basic demographic information and a series of questions on health insurance coverage, but before the more detailed questions on recent illness and use of health services. The precise questions asked were:

Question 7. 0
In general, would you say that your health is excellent, good, fair or poor?
Question 7. 1
I now want to ask you about how you feel generally. Overall, would you say that you're very happy, happy, unhappy or very unhappy?

Table 3.2 provides a cross-tabulation of the responses to these the two questions. The results show that, although the respondents generally described themselves as being in good (or excellent) health and happy, the overlap between those who were not happy and those in poor health was not great. Thus, of the 4.5% of respondents who assessed themselves as being either 'unhappy' or 'very unhappy', only 1.5% or a third also assessed their health as being 'poor'. In contrast, of the 4.9% who assessed their health as being 'poor', less than one third (1.5%) assessed themselves as either 'unhappy' or 'very unhappy'. Almost 70% of those who assessed their health as being 'poor' were either 'very happy' or 'happy', while more than 30% of those who were 'very unhappy' assessed their health status as being either 'excellent' or 'good'. Clearly, poor health (once accepted and adjusted to) does not prevent happiness, any more than good health guarantees it.

When those individuals who reported themselves as being either 'unhappy' or 'very unhappy' were compared according to whether or not they belonged to families who were above or below the poverty threshold described above, considerable differences were found between those of different age and in different family types. The degree of unhappiness amongst those below the poverty threshold was almost two and a half times higher than amongst the rest of the population, with a differential of about this magnitude apparent for most family types. Furthermore many of these differences were statistically significant, even at the 0.01 level (Saunders, 1996, Table 4.3). When the differential pattern of responses to the subjective health status question was analysed, the differ-

Table 3.2: Cross-tabulation of Individual Self-assessments of Happiness and Health (Percentages)

| | | Self-assessed happiness | | | | |
		'very happy'	'happy'	'unhappy'	'very unhappy'	Total
	'Excellent'	13.4	14.5	0.3	0.1	28.3
Self-assessed	'Good'	11.0	37.9	1.1	0.1	50.1
health	'Fair'	2.0	13.4	1.3	0.1	16.8
status	'Poor'	0.4	3.0	1.2	0.3	4.9
	Total	26.8	68.7	3.9	0.6	100.0

Source: Saunders, 1996, Table 4.2.

ences between those reporting poor health who were either side of the poverty threshold were again significant in many cases (Saunders, 1996, Table 4.4).

Overall, these results thus provide support for the view that those whose incomes are below the poverty line, or only marginally above it, perceive themselves as both more unhappy and in worse health than other Australians. These associations hold at both the aggregate level and also for most of the family (or income unit) types typically studied in poverty research. However, it is also clear that there are several other factors, specifically but not exclusively age, which also influence the differences revealed in these results.

Poverty and Stress

Information on a vast range of reported medical conditions was collected in the course of the NHS, although it should be noted that the reliability of this information was not verified against independent medical advice. Examining variations in the incidence of these conditions amongst the survey population provides the basis for a more rigorous examination of the linkages between economic status and health status.

In choosing which of the hundreds of available variables to consider for further analysis, the starting point was the arguments developed by Wilkinson and other writers concerning how the lack of adequate economic resources and sense of deprivation associated with poverty can lead to situations of psychological stress and produce a loss of self-esteem. That loss, when combined with the emotional and psychological stresses associated with the constant struggle to make ends meet can explain how poverty can lead to the kinds of stress which are associated with poor health.

Information about the incidence of four specific stress-related conditions which related to this form of stress was collected in the NHS. The four conditions were:

- nerves, tension, emotional problems, etc.;
- insomnia;
- headache – due to stress/tension; and
- depression.

In each case, the variable refers to whether or not each condition was experienced in the two-week period prior to the survey interview.

The incidence of these four stress-related conditions was combined into a single measure reflecting the incidence of experiencing *one or more* of them. (Analysis of those reporting *more than one condition* was not viable because of the very small numbers involved.) Figure 3.5 shows how the incidence of the composite stress variable varies by family type and poverty status.

These results imply, for example, that the incidence of stress as measured here is around 9% for non-poor couples with children, but almost 11% for poor couples with children. For poor sole parents with children, the incidence is over 18%. Thus, if a (non-poor) couple with children were to separate, the custodial parent (normally the mother) would find, if she were to fall into poverty, that on average the probability of experiencing stress-related ill-health would more than double from below 9% before separation to over 18% afterwards. In all cases except for aged couples and sole parents, the differences in the incidence of stress at this level of analysis are statistically significant. To become poor is to experience not only financial stress but also to confront a greatly-increased risk of stress-related ill-health.

It is, of course, possible that the significant differences revealed in the above results are either spurious or contingent, in that they reflect the role of intervening variables. In order to assess the likelihood of these explanations, the same data were used in a multivariate analysis using the technique of logistic regression. The independent variables included in the

Figure 3.5: The Incidence of One or More Stress-Related Conditions by Income Unit Type and Poverty Status[a]

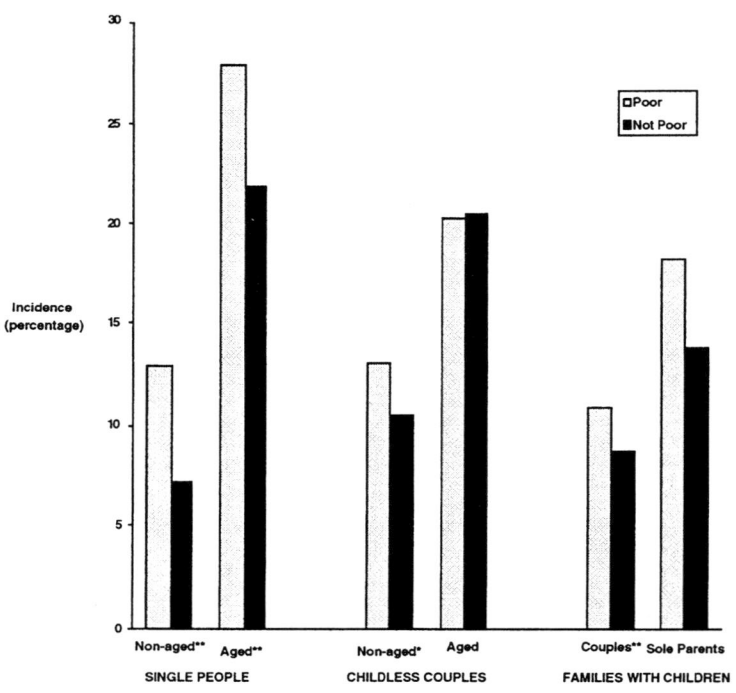

Note: a) The asterisks (*/**) indicate that the differences by poverty status are significant at the (0.05/0.01) level.

model were the six family types used earlier; a three-way dummy variable describing each individual's labour force status as either employed, unemployed or not in the labour force; and further dummy variables reflecting whether or not the respondent was female, was a smoker at the time of the survey, and whether or not the individual's family was above or below the 120% poverty line threshold.

The estimated parameters on virtually all of the independent variables in the model were statistically significant. Of particular importance, the impact of poverty status on stress remained significant, even after controlling for the influence of family type (and thus age), labour force status, gender and smoker status. The results in Table 3.3 indicate how the probability of experiencing one or more of the four stress-related conditions varies with the socioeconomic characteristics of each individual.

The individual entries shown in Table 3.3 imply, for example, that a single, non-smoking, non-aged, employed male above the poverty threshold has, on average, a 6% probability of experiencing one or more stress-related conditions. If the same individual were to become unemployed, the probability would rise very slightly. If (in the more likely scenario) they were to become unemployed and move from being above to below the poverty threshold, the probability of experiencing stress would rise more significantly, to 8%. If the person was a smoker, those probabilities would all be slightly higher. If the person was female (and a smoker), the probabilities would be a further 5 percentage points or so higher again.

The main focus of interest in these results is on how poverty status affects stress. It can be seen that in the great majority of cases, those below the poverty threshold have a stress probability between 1 and 3 percentage points above those in the same circumstances who are above the poverty threshold. While this differential may not appear that large, in some cases a 2 percentage point increase in the stress probability is substantial relative to the initial probability level.

A different perspective on the size of the poverty effect can be obtained by comparing the hypothetical situation for an employed (non-smoking) non-poor, married mother with that of a non-smoking female sole parent who is poor and not in the labour force. The respective stress probabilities in these two cases are 0.10% and 0.18% – a proportionate increase in the likelihood of experiencing stress of 80%. These comparisons illustrate that the size of the effects implied by the results in Table 3.3 are substantial – particularly when it is borne in mind that many of the effects which are estimated separately in the model often occur simultaneously, leading to even bigger cumulative effects on health.

The main conclusion to be drawn from these estimates is that the impact of poverty on stress is not a statistical artefact resulting from the effects of other variables like income unit type, gender or labour force status being wrongly ascribed to an apparent poverty affect. The correlation between poverty and stress is a real one. Poverty exerts an *independent* effect on stress, the size of which is substantial when compared with the impact of other stressful life experiences such as becoming unemployed or separation in the case of those with children.

Table 3.3: Logistic Regression Model Prediction Probabilities for Adult Individuals Experiencing Stress-related Illness[a]

	Single, non-aged		Single, aged	Non-aged couple		Aged couple	Couples with children		Sole parent
	employed	unemployed	NILF	employed	unemployed	NILF	employed	unemployed	NILF
(a) MALES									
Non-smoker									
Above poverty threshold	0.06	0.06	0.12	0.06	0.07	0.12	0.05	0.06	0.09
Below poverty threshold	0.07	0.08	0.15	0.08	0.09	0.15	0.07	0.07	0.10
Smoker									
Above poverty threshold	0.06	0.07	0.14	0.07	0.08	0.14	0.06	0.07	0.10
Below poverty threshold	0.08	0.09	0.16	0.09	0.10	0.16	0.08	0.08	0.12
(b) FEMALES									
Non-smoker									
Above poverty threshold	0.10	0.11	0.20	0.11	0.12	0.20	0.10	0.11	0.15
Below poverty threshold	0.12	0.13	0.24	0.14	0.15	0.24	0.12	0.13	0.18
Smoker									
Above poverty threshold	0.11	0.12	0.23	0.13	0.14	0.23	0.11	0.12	0.17
Below poverty threshold	0.14	0.15	0.27	0.15	0.17	0.27	0.13	0.15	0.20

Income unit type/labour force status[b]

Notes: (a) Predicted probabilities derived from the estimates of the logistic regression model shown in the final column of Saunders, 1996, Table 4.7.

(b) NILF = not in the labour force.

Source: Saunders, 1996, Table 4.8.

Discussion and Social Policy Implications

The results presented in the previous section cannot be claimed to substantiate the view that relative poverty in Australia exerts an influence on stress-related poor health which is *independent* of the direct impact of poverty on the health. As noted earlier, this question cannot be answered using cross-section data for a single year in one country. Nevertheless, the results do indicate that poverty has adverse consequences for health even when an explicit relative standard is used to measure poverty.

From this perspective, the Australian results are consistent with Wilkinson's thesis that inequality causes a decline in health via its effects on stress, self-esteem and related psycho-social influences. They also suggest that any increase in the incomes of the rich will adversely impact upon the overall level of health because they will cause the poverty line to rise and the numbers in poverty along with it, although this effect will be extremely weak and difficult to discern.

What does all this imply for social policy? The first and most obvious implication of the results described above is that if it can be demonstrated that increasing inequality has adverse consequences for health which result from changes in *relative* income, then the case for redistributing income is strengthened. Even if such a case cannot be made, the fact that this line of research has highlighted the impact of low income relative to the mean (or median) on health status provides a basis for broadening the context within which health policy is conceptualised, as well as providing further evidence that poverty is bad, not only in and of itself, but also because it has consequences which are themselves bad.

However, in light of the extreme frailty of much of the empirical research linking inequality in the distribution of income to health, considerable caution must be applied before any policy conclusions can be deduced. It seems on the face of it inherently implausible, for example, that increasing income taxes on the rich will improve the health status of the poor, although it is possible that those countries (or states or provinces) that are more prepared to impose higher taxes on the rich may also provide better services and supports which will raise the health standards of poor and rich alike. Trying to uncover whether this is the kind of process underlying the observed statistical correlations should be an important component of the future research agenda.

Until more convincing evidence on causal mechanisms is produced, the policy implications of what is currently known remain unclear. If social policy is to turn back the tide of economic liberalism, it will need to do so from a position based upon better research. Achieving this will involve greater openness in the specification of research questions, more rigour in the design of research studies and improvement in the application of logic and objectivity in deriving inferences (for policy, behaviour and causes) from research findings.

The failure of the social policy research community to be sufficiently professional in much of its work has, in my view, assisted the economic liberalists to take over the field with their own studies, often based on the application of simplistic neoclassical economic models of rational choice which are devoid of institutional understanding and policy detail. Turning back the liberalist tide will involve challenging the values which underlie it, but it will also require an intellectual attack on the methodology used in these studies and the inferences drawn from them.

In light of the current juncture of the academic debate on many of these issues, it is premature to claim too much from the existing literature regarding the causative links between inequality of income and the level of, or inequalities in, health. Aggregate correlations alone provide no basis for policy intervention unless the implied relationships are stable over time and place, are relatively insensitive to variations in measurement and technique, and can be shown to be underpinned by well-specified and coherent pathways of influence which have themselves been confirmed by empirical inquiry and are based upon theorised models of choice, behaviour and structure.

In my view, none of these conditions can be said to prevail currently in relation to the impact of income distribution on health. Even if it were accepted that there is such a relation, what does it imply for the design of redistributive policies? Can we be certain, for example, that *any* policy which redistributes income will improve health? Surely, the precise *form* of such policies will matter, including whether they are achieved by increasing taxes on the rich, by raising transfer incomes of the poor, or by increasing wages at the expense of profits, to name but a few possibilities. Whilst each of these measures can be designed so as to have a similar initial impact on inequality, each will result in different flow-on effects and have different economic consequences, which will influence the final redistributive outcome. It seems implausible to believe that these differences will not themselves lead eventually to differential consequences for health.

Economists have long argued that the incentive effects of taxes and transfers on labour supply and savings have the potential to distort the impact of social programmes and undermine their purpose. They are right, in principle at least, if not necessarily in practice. In order to decide how important these effects are, they need to be analysed within a long-term behavioural framework. Even though the immediate consequences for the income of an employed person from accepting a low-wage job may suggest that continued receipt of unemployment benefit may be the preferred option, any form of waged employment provides direct access to the world of work and thus to opportunities for better-paid jobs in the longer term. If what matters for behaviour, and thus for self-esteem and a sense of self-worth, is not so much today's income, but rather the prospects for tomorrow, then this needs to be incorporated into studies of how income and inequality affect health.

It is also worth emphasising that the existence of a relationship between income and health has potentially important consequences, not only for policies which seek to redistribute income, but also for the funding and operation of the health care system. At the very least, attention needs to be given to the socioeconomic determinants of health in the context of allocating health resources as well as in the broader context of allocating *overall* resources between the health sector and other competing uses. We may not yet be in the situation described by Wilkinson as one in which the policies introduced by the Ministers responsible for taxation or social security may have a greater influence on health than those introduced by the Minister of Health, but there is an element of truth in such assertions which should be heeded.

Summary
Neither Australia nor New Zealand has been protected from the widespread trend to inequality which has been taking place since around the mid-1970s. Gone are the days when study on income distribution was about as exciting as watching the grass grow. Change is now *the* feature of income distribution in most countries, though generally not the kind of change which is welcomed by those with egalitarian leanings. The fact that the trend to inequality has generally been market-driven and has shown up most clearly in the distribution of market incomes means that increasing pressure is being placed on redistributive mechanisms.

At the same time, fiscal imperatives are reducing the resources available for redistributive programmes on the expenditure side, while politicians still court the electorate with the promise of further cuts in taxation. The Australasian response to the squeeze that this has placed on social programmes has been to make them increasingly targeted. However, there are limits as to how far this process can go, both as a consequence of the increased disincentive effects which accompany income targeting, as well as from the design and operational complexities associated with programmes which target through other means.

In light of the current and future prospects for progressive social reform in both countries, we certainly need more arguments for equality. But we also need *good* arguments, in the sense that they have a solid foundation and are amenable to wide application. Although it is tempting to count the argument that inequality is bad because it exerts an undesirable and *independent* influence on health as one of these, the argument and analysis presented in this paper suggests that this is premature in terms of scholarship and unwise as a matter of strategy.

Although this paper has been critical of aspects of the work published by Wilkinson and others on the links between income inequality and health, this should not be interpreted as implying criticism of the larger project on which these scholars are engaged. Rather, its aim has been to attempt to strengthen the research basis of that work by being critical in a constructive way. Nothing that has been written here should detract attention from the issues being addressed as part of the broader project.

Much has already been written on the measurement of inequality; it is time to turn towards investigating the consequences of inequality, as I am sure those in New Zealand need no reminding. Will the argument that economic inequality produces poorer health outcomes provide the basis for a new egalitarianism? Possibly, but not yet.

References

Atkinson, AB, L Rainwater and TM Smeeding (1995) *Income Distribution in OECD Countries: The Evidence from the Luxembourg Income Study*, Social Policy Studies No. 18, Paris: OECD

Bishop, JA, JP Formby and WJ Smith (1991) 'International comparisons of income inequality: tests for Lorenz dominance across nine countries', *Economica*, 58, pp 461-77

Commission of Inquiry into Poverty (1975) *First Main Report. Poverty in Australia*, Canberra: AGPS

Cowell, FA (1995) *Measuring Inequality*, 2nd edn, London: Harvester Wheatsheaf

Forster, DP (1992), 'Income distribution and life expectancy: correspondence', *British Medical Journal*, 34, March, pp 715-6

Gruen, FH (1995) 'The Australian welfare state: neither egalitarian saviour nor economic millstone?', *Economic and Industrial Relations Review*, 6 (1), pp 125-38

Harding, A and D Mitchell (1992) 'The efficiency and effectiveness of the tax-transfer system in the 1980s', *Australian Tax Forum*, 9 (3), pp 277-304

Jenkins, SP (1991) 'The measurement of income inequality', in L Osberg (ed.) *Economic Inequality and Poverty. International Perspectives*, New York: ME Sharpe, pp 3-38

Judge, K (1995) 'Income distribution and life expectancy: a critical appraisal', *British Medical Journal*, 311, November, pp 1282-85

Kaplan, GA et al (1996) 'Inequality in income and mortality in the United States: analysis of mortality and potential pathways', *British Medical Journal*, 312, April, pp 999-1003

Kennedy, BP, I Kawachi and D Prothrow-Stith (1996) 'Income distribution and mortality: cross-sectional ecological study of the Robin Hood Index in the United States', *British Medical Journal*, 312, pp 1004-07

Le Grand, J (1987), 'Inequalities in health: some international comparisons', *European Economics Review*, 31, pp 182-91

Lundberg, O and J Fritzell (1994) 'Income distribution, income change and health: on the importance of absolute and relative income for health status in Sweden', in LS Levin, L McMahon and E Ziglio (eds) *Economic Change, Social Welfare and Health in Europe*, Copenhagen: World Health Organisation, pp 37-58

McIsaac, SJ and RG Wilkinson (1995) *Cause of Death, Income Distribution and Problems of Response Rates*, Working Paper No. 136, Luxembourg: Luxembourg Income Study

Rodgers, GB (1979) 'Income and inequality as determinants of mortality: an international cross-section analysis', *Population Studies*, 33 (2), pp 343-51

Saunders, P (1996) *Poverty, Income Distribution and Health: An Australian Study*, SPRC Reports and Proceedings No. 128, Social Policy Research Centre, University of New South Wales

Smeeding, T (1996) 'America's income inequality: where do we stand?',

Challenge, September-October, pp 45-53

Smith, GD (1996) 'Income inequality and mortality: why are they related?', *British Medical Journal,* 312, April, pp 987-8

Torrey, B, T Smeeding and D Bailey (1996) *Rowing Between Scylla and Charybdis? Income Transitions in Central European Households,* Working Paper No. 132, Luxembourg: Luxembourg Income Study

Wilkinson, RG (1992) 'Income distribution and life expectancy', *British Medical Journal,* 34, January, pp 165-168

Wilkinson, RG (1993) 'Income and health', in *Health, Wealth and Poverty. Papers on Inequalities in Income and Health,* London: Medical World Publications, pp 6-11

Wilkinson, RG (1994) 'Health, redistribution and growth', in A Glyn and D Miliband (eds) *Paying for Inequality. The Economic Cost of Social Injustice,* London: Rivers Oran Press, pp 24-43

Wilkinson, RG (1995) 'Commentary: a reply to Ken Judge: mistaken criticisms ignore overwhelming evidence', *British Medical Journal,* 311, November, pp 1285-7

World Bank (1993) *World Development Report 1993. Investing in Health,* Washington, DC: Oxford University Press

4 ~ Income Distribution, Social Capital and Mortality

Ichiro Kawachi and Bruce P Kennedy

Introduction

Inequalities in health by socioeconomic status are large, pervasive, persistent and widening over time (see Pappas et al, 1993; Pearce, Marshall and Borman, 1991). The proposed mechanisms for the SES gradient in health include social selection (the 'drift' hypothesis), unequal access to health care, differences in lifestyle, as well as material deprivation and exposures to different conditions of living, such as housing and work stress (see Townsend, Davidson and Whitebread, 1992). A comparatively recent – and less well understood – hypothesis posits that the unequal *distribution* of income matters for health as much as the absolute level of income (see Wilkinson, 1992 and 1996). In contrast to existing research on social inequalities in health – which uses *individual*-level variables such as income, educational attainment or occupational ranking as the measure of SES – the new approach focuses on the societal distribution of income (a *macro*-level variable) as the determinant of health.

Given the marked polarisation of income during the past decade in New Zealand, Britain, the USA and many other countries, this new perspective may be particularly relevant for understanding contemporary trends in population health in these societies. The aim of this paper is to review the theory and evidence linking income distribution to mortality, and to propose some mechanisms by which political economy affects health.

The Relationship of Income Distribution to Life Expectancy and Mortality

Almost every study of SES and health has concluded that as we descend the hierarchy of income, rates of illhealth increase (see Figure 4.1) (see Kawachi et al, 1994). Although this paper will focus on mortality and life expectancy, the income gradient itself has been replicated using virtually every measure of health outcome, including morbidity, disability and

Figure 4.1: Relationship of Income to Life Expectancy

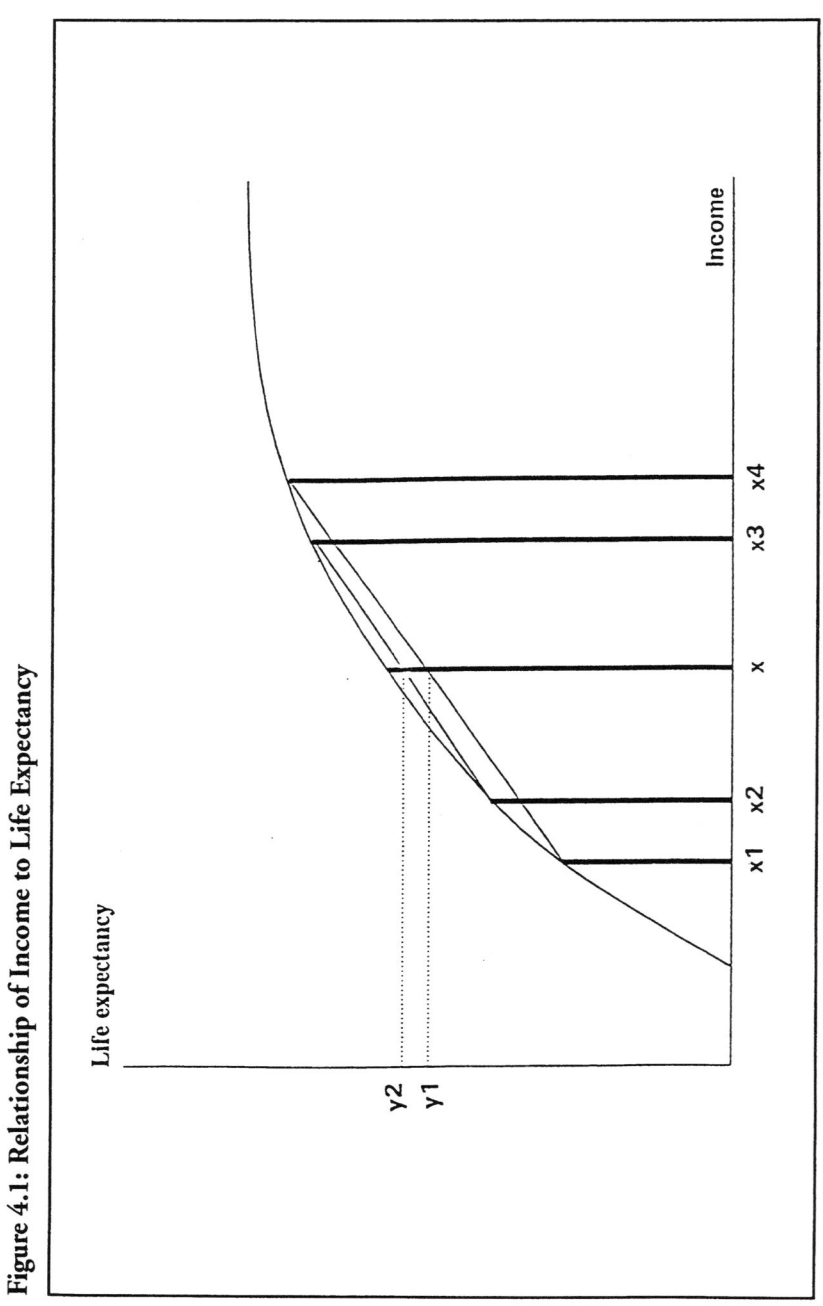

perceived health status. One of the universally observed characteristics of the income/life expectancy (or income/mortality) curve is that its slope declines with increasing income, i.e., there are diminishing returns to rising income (see Figure 4.1). In other words, the income/life expectancy curve is steep in the regions of absolute income deprivation; but it levels off beyond a certain standard of living. This characteristic non-linear relationship appears to hold true not only for data within a single country, but also for comparison of data across different countries (see Wilkinson, 1996; Kawachi et al, 1994).

One important consequence of the shape of the income/life expectancy curve is that the *distribution* of income must influence the average life expectancy of a country. The tendency for greater dispersion of income to be associated with lower mean life expectancy can be demonstrated by considering a hypothetical society with mean income x in Figure 4.1. If the income dispersion is between x_1 and x_4, mean life expectancy would be y_1 (assuming equal numbers of people on either side of the mean income). Now if income dispersion in this society is reduced, by taking $(x_4$ minus $x_3)$ and transferring that amount to raise the incomes of the less well-off, from x_1 to x_2, then *ceteris paribus*, mean life expectancy in this society would rise to y_2. In other words, redistributing income to the poor may raise overall life expectancy, even if the average level of income remains unchanged. This prediction is the consequence of the downwardly concave relation between income and life expectancy, i.e., the rise in life expectancy among the poor (as a result of redistribution) more than offsets any loss in life expectancy among the rich (see Rodgers, 1979). Following the same line of reasoning, we might expect that two countries with the same average income but different income distributions would experience different levels of health, the country with the more equitable income distribution having a higher average life expectancy than the country with less equitable income distribution.

In fact, a growing number of studies have found such a relationship. The association between income distribution and mortality was first reported by Rodgers (1979) using data from 56 developed and less developed countries in 1965. In recent years, the hypothesis has become most closely identified with the work of Richard Wilkinson (1986, 1990, 1992, 1996) who reported the income/life expectancy association in a series of studies based on OECD country data. Although there has been some debate over the robustness of the evidence based on OECD data (see

Judge, 1995; also Saunders, chapter 3), several other international studies have also corroborated the link between income distribution and mortality, particularly infant mortality (see Flegg, 1982; Waldmann, 1992; Wennemo, 1993). Without question, these cross-national comparisons of income distribution and health are fraught with difficulties, including considerations of purchasing power parity as well as confounding factors such as culture.

In 1995, however, two independent studies were published in the *British Medical Journal* which provided for the first time a test of the income inequality hypothesis within a single country, the United States (see Kennedy, Kawachi and Prothrow-Stith, 1996; Kaplan et al, 1996). Kennedy et al examined the relationship between the degree of household income inequality across the 50 US states and state-level variation in all-cause and cause-specific mortality. The degree of income inequality in each state was estimated by the Robin Hood Index, which is equivalent to the proportion of aggregate income that must be redistributed from households above the mean, and transferred to those below the mean in order to achieve an equal distribution of household incomes.

The higher the Robin Hood Index, the more unequal is the distribution of income. The Robin Hood Index for the United States overall was 30.2% in 1990. The overall correlation of the Robin Hood Index to age-standardised total mortality in 1990 was 0.54 (p < 0.0001). After adjusting for poverty and median income, a 1% rise in the Robin Hood Index was associated with an increase in age-adjusted total mortality rate of 21.7 deaths per 100,000 (95% confidence interval: 6.6 to 36.7). The Robin Hood Index was also associated with deaths from coronary heart disease, neoplasms, homicide, infant mortality, as well as the so-called causes of death 'amenable' to medical intervention (see Kennedy et al, 1996).

In the same issue of the *British Medical Journal*, Kaplan et al reported a similar association between mortality and income inequality, again using US data. Their measure of income inequality was the proportion of total income earned by the bottom 50% of households in each state. The income distribution/mortality association in both US studies appeared to be robust – it was present in men and women, and in blacks as well as whites, and appeared to be independent of median income, urban/rural residence and even cigarette smoking. The association persisted even after adjusting household incomes for taxes and transfer payments, as well as differences in household size using equivalence scales (see Judge, 1996).

Finally, the choice of income distribution measure did not appear to influence the relationship; no matter what measure was used – the Gini coefficient, the Robin Hood Index, the Atkinson Index, the decile ratio and so on – all were strongly correlated with mortality (see Kawachi and Kennedy, submitted).

Mechanisms Relating Income Distribution to Life Expectancy and Mortality

As yet little is understood of the mechanisms that link income inequality to life expectancy and mortality. Two potential explanations (among several) are that:

(a) the relationship simply reflects the effect of poverty on health; and
(b) a wide gap in income leads to a change in exogenous factors (e.g., disinvestment in human and social capital) that shifts the income/life expectancy curve inwards.

These two potential mechanisms – not necessarily mutually exclusive – are discussed in turn.

The Effect of Poverty

The relationship of poverty to poor health is well described (see Townsend et al, 1992). In the US data, rates of poverty in each state (as measured by the proportion of households falling below the official US poverty threshold) were moderately correlated with mortality rates ($r = 0.42$). In turn, states with the greatest income inequality, as measured by the Robin Hood Index, also tended to have the highest rates of poverty ($r = 0.73$) (see Kennedy et al, 1996; Kawachi and Kennedy, submitted). In order to separate the effects of poverty on mortality from the effects of income distribution, two analytical approaches have been tried: simultaneously adjusting for poverty and income distribution in multivariate regression models, and alternatively, examining the income distribution/mortality link *within strata* of high and low-poverty regions in the USA (see Kennedy et al, 1996; Kawachi and Kennedy, submitted). Under both approaches, income distribution remained a statistically significant predictor of mortality, while the effect of poverty was often attenuated. Thus, while the deleterious effect of poverty on ill health is well-established (and in many ways much better documented), there also appears to be an independent effect of income distribution on mortality. Although the eradication of poverty remains an important public health goal, it may not be sufficient to eliminate the excess burden of ill health resulting from relative deprivation.

Shifts in the Income/Life Expectancy Curve

A further mechanism by which income inequality may lead to excess deaths is through inward shifts of the income/life expectancy curve (Figure 4.2). In other words, we posit that changes in the underlying distribution of income may induce changes in factors that are exogenous to the income/mortality relationship. For example, worsening income inequality may lead to disinvestment in human capital and social capital, with the result that society is less capable of translating a given increase in income ($X - X'$) into health improvements (point B instead of point A').

There is as yet limited evidence to suggest that societies which tolerate wide income disparities also tend to under-invest in human capital. Kaplan et al (1996) examined this hypothesis in the US. They reported that states with greater income inequality also had higher rates of high-school dropouts ($r = 0.50$), lower rates of 4th grade reading ($r = -0.58$) and math proficiency ($r = -0.64$), lower educational spending as a proportion of the total state budget ($r = -0.32$), and fewer library books per capita ($r = -0.42$). Although preliminary, these data support the notion that wide income disparities may generate pressures to disinvest in the public education system through two mechanisms: exit of higher income groups into the private education system; and secondarily, political pressure from the same groups to cut taxes that finance public education (which no longer support the education of the well-off).

Income Distribution, Social Capital and Mortality

In a seminal essay on the dysfunctions of social stratification published in 1953, Melvin Tumin speculated that, "to the extent that inequalities in social rewards cannot be made fully acceptable to the less privileged in a society, social stratification systems function to encourage hostility, suspicion and distrust among the various segments of society and thus to limit the possibilities of extensive social integration".

In the last five years, there has been a tremendous surge in interest in the notion that social cohesion, or social capital, is a crucial determinant

Figure 4.2: Hypothetical Shift in Income/Life Expectancy Curve

Shift in income/life expectancy curve

Life expectancy

A'

A

B

X

X'

Income

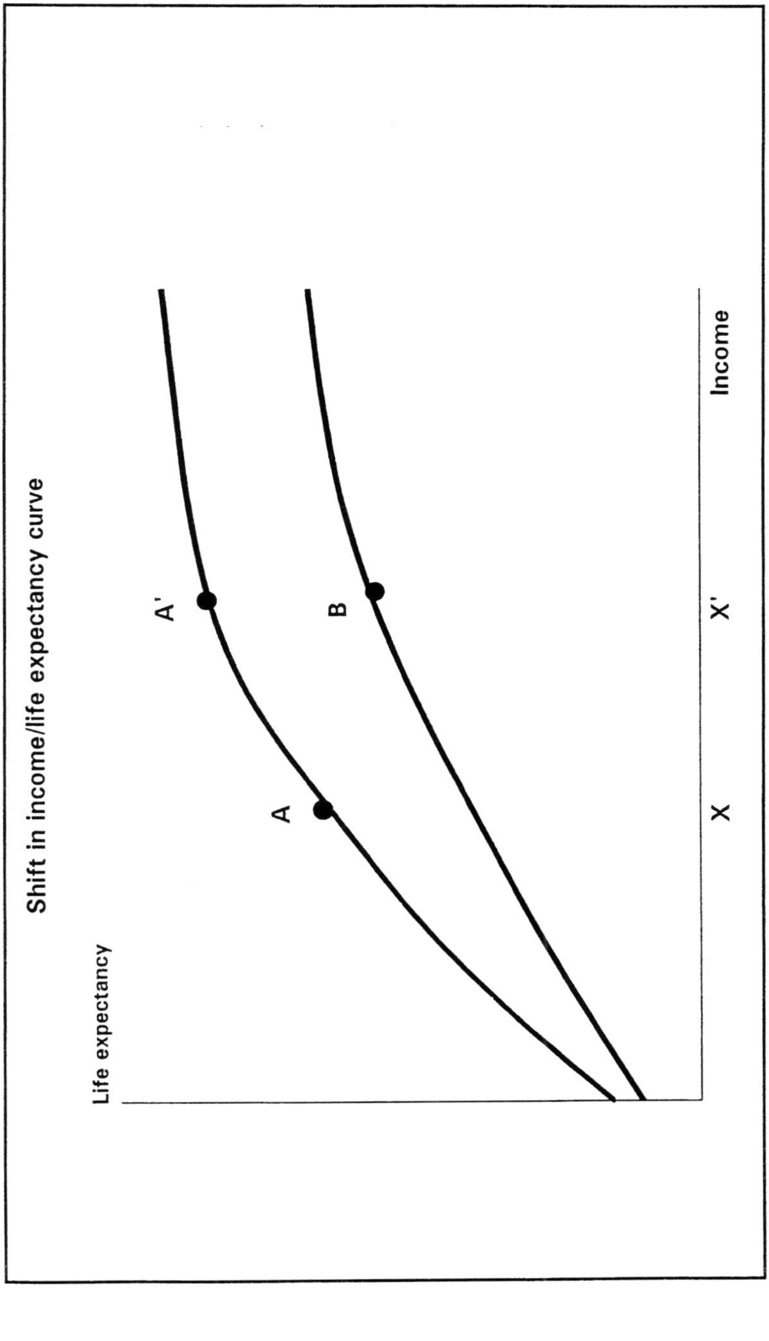

of the functioning of societies and of the health of its members. Social capital, as defined by its principal theorists (see Putnam, 1993 and 1995; Coleman, 1990), refers to those features of social organisation, such as civic participation, norms of reciprocity and trust in others, that facilitate cooperation for mutual benefit. The level of social capital in a society has been claimed to be associated with several benefits, including the better functioning of the civic institutions of democracy, as well as the fostering of economic development (see Putnam, 1993). Social capital may be conceptualised as a societal-level (or *ecologic*) characteristic, whose counterpart at the individual level is measured by a person's connection to social networks and sources of support. In the public health realm, a vast literature has linked improved health outcomes (lower morbidity and mortality) to an individual's level of social connectedness (see Berkman and Syme, 1979; House, Landis and Umberson, 1988; Kawachi et al, 1996a).

In an attempt to test the linkages between income inequality and breakdown of social cohesion, Kawachi et al (1996b) examined the cross-sectional ecological relationships between income distribution and indicators of social capital within the United States. Income inequality was measured by the Robin Hood Index. Following the work of Putnam (1993), social capital was measured by two indicators – level of civic engagement (per capita membership in community and voluntary groups) and levels of mutual trust among community members – obtained from weighted responses to the General Social Survey, conducted by the National Opinion Research Center (see Kawachi et al, 1996b). Data were available in 39 US states.

As hypothesised, income inequality was strongly associated with reduced levels of social capital. States with greater income inequality had lower levels of per capita membership in groups, such as church groups, labour unions, sports groups, professional or academic societies, school groups, political groups and fraternal organisations ($r = -0.4$, $p < 0.01$). Residents in states with greater income inequality also reported lower levels of social trust, as assessed by the proportions of respondents who agreed that "most people cannot be trusted" ($r = 0.73$, $p < 0.0001$). In turn, both indicators of social capital were strongly associated with mortality rates – the greater the extent of disinvestment in social capital, the higher the level of mortality (Figure 4.3). These data are thus consistent with the notion that income inequality may lead to poor health outcomes via disinvestment in social capital (Kawachi et al, 1996b).

Figure 4.3: Relationship of Social Trust to Mortality in the USA

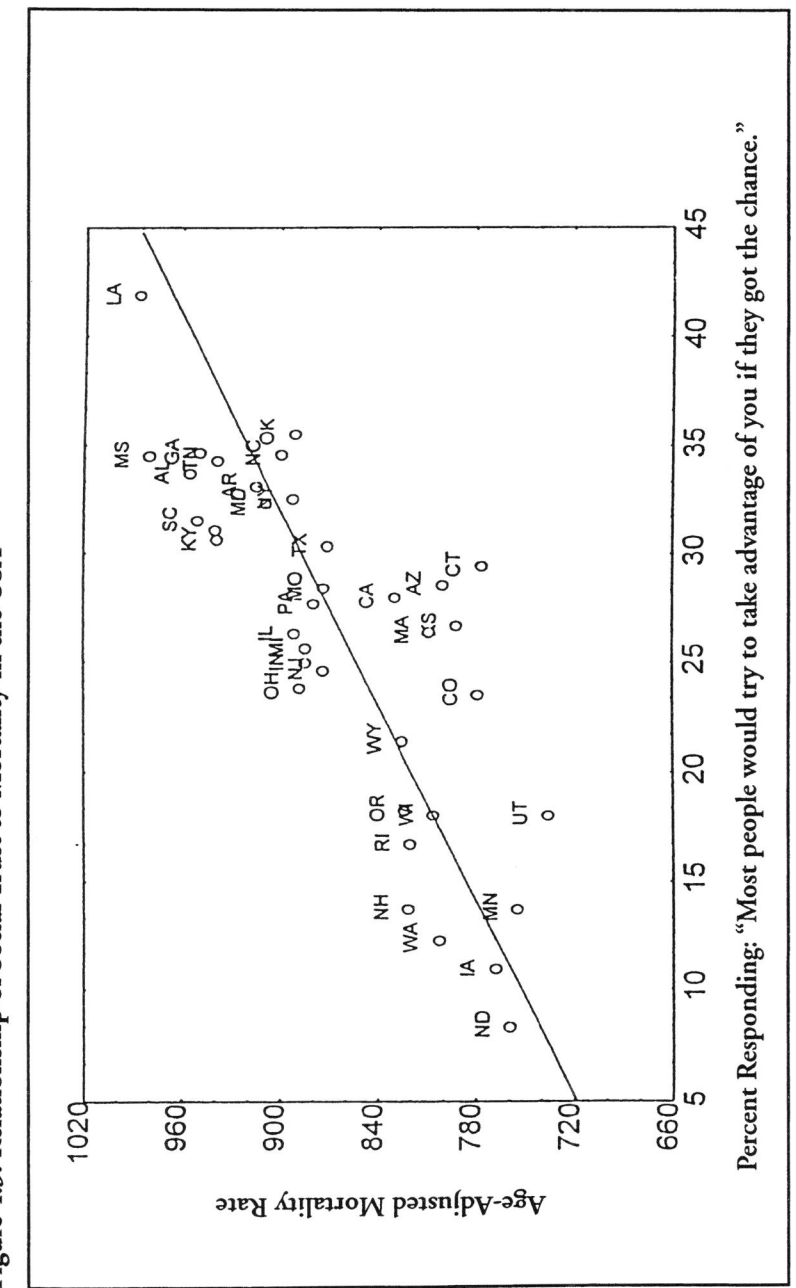

Percent Responding: "Most people would try to take advantage of you if they got the chance."

What Do We Need to Know?

The relationship between income inequality and mortality has been reported by a sufficient number of different investigators (using different data and methods) to indicate that it merits serious attention, in particular, to improve on the limitations of existing studies. Apart from the obvious improvements which could be made – such as conducting longitudinal time series (as opposed to cross-sectional) analysis, and extending the range of health outcomes to include measures of morbidity as well as quality of life – there are two issues that could be resolved only by the use of better data:

(a) at what level of areal aggregation does income inequality exert its effect on mortality?; and

(b) does income inequality predict *individual* risk of mortality?

Answering these two questions calls for the simultaneous collection of data at multiple levels – societal (as in cross-national comparisons), state-level, census-tract level, neighbourhood level and individual level. The analysis of this type of data also calls for types of methods – such as random effects mixed-level regressions – that are different from those encountered in conventional epidemiological research (e.g., multivariate logistic regression). Pinpointing the level of aggregation at which income distribution affects health outcomes should provide important clues about the aetiological pathways involved. For example, if income inequality acts on health through the passage of deleterious social policies (e.g., reductions in the social safety net, or disinvestment in public education), then we might expect to detect health effects at politically-meaningful units of aggregation, such as states or entire nations. Alternatively, if income inequality harms health through effects at the neighbourhood level (e.g., via residential segregation and reduced choice of healthy foods in neighbourhood groceries), then one would expect to detect links to health at these levels. Similarly, demonstrating the link between income distribution (an obligatory ecological variable) and *individual* health outcomes is of salient aetiological interest, since we need to know who is most vulnerable to the effects of income inequality – whether they are the poor, the elderly, ethnic minorities, women or children.

Conceptualising and analysing the effects of *location* on health – as opposed to analysing the contribution of personal characteristics – lies at the frontier of research on socioeconomic inequalities in health (see MacIntyre, McIver and Soloman, 1993). Thinking about income distribution encourages us to confront the ultimate societal influences on public health.

References

Berkman, LF and SL Syme (1979) 'Social networks, host resistance and mortality: a nine-year follow-up study of Alameda County residents', *American Journal of Epidemiology*, 109, pp 186-204

Coleman, JS (1990) 'Social capital' (chapter 12, pp 300-321), in JS Coleman *Foundations of Social Theory*, Cambridge: Harvard University Press

Flegg, A (1982) 'Inequality of income, illiteracy, and medical care as determinants of infant mortality in developing countries', *Population Studies*, 36, pp 441-58

House, JS, KR Landis and D Umberson (1988) 'Social relationships and health', *Science*, 214, pp 540-45

Judge K (1995) 'Income distribution and life expectancy: a critical appraisal', *British Medical Journal*, 311, pp 1282-5

Judge K (1996) 'Income and mortality in the United States (letter)', *British Medical Journal*, 313, pp 1206-7; *see also* replies by Kaplan et al, and Kennedy, Kawachi and Prothrow-Stith (p 1207)

Kaplan, G et al (1996) 'Inequality in income and mortality in the United States: analysis of mortality and potential pathways', *British Medical Journal*, 312, pp 999-1003

Kawachi, I et al (1994) *Income Inequality and Life Expectancy – Theory, Research and Policy*, Boston: The Health Institute, New England Medical Center

Kawachi, I et al (1996a) 'A prospective study of social networks in relation to total mortality and cardiovascular disease in men in the US', *Journal of Epidemiology and Community Health*, 50, pp 245-51

Kawachi, I et al (1996b) 'Social capital, income inequality, and mortality', *American Journal of Public Health*

Kawachi, I and BP Kennedy (submitted) 'The relationship of income inequality to mortality – does the choice of indicator matter?'

Kennedy, BP, I Kawachi and D Prothrow-Stith (1996) 'Income distribution and mortality: cross-sectional ecological study of the Robin Hood Index in the United States', *British Medical Journal*, 312, pp 1004-07

MacIntyre, S, S McIver and A Soloman (1993) 'Area, class and health: should we be focusing on places or people?', *Journal of Social Policy*, 22, pp 213-34

Pappas, G et al (1993) 'The increasing disparity in mortality between socioeconomic groups in the United States, 1960 and 1986', *New England Journal of Medicine*, 329, pp 103-9

Pearce, NE, S Marshall and B Borman (1991) 'Undiminished social class mortality differences in New Zealand men', *New Zealand Medical Journal*, 104, pp 153-6

Putnam, RD (1993) 'Tracing the roots of the civic community' (chapter 5, pp 121-162), in RD Putnam *Making Democracy Work. Civic Traditions in Modern Italy*, Princeton, NJ: Princeton University Press

Putnam, RD (1995) 'Bowling alone: America's declining social capital', *Journal of Democracy*, 6, pp 65-78

Rodgers, GB (1979) 'Income and inequality as determinants of mortality: an international cross-section analysis', *Population Studies*, 33, pp 343-51

Townsend, P, N Davidson and M Whitehead (eds) (1992) *Inequalities in Health: The Black Report and the Health Divide*, Harmondsworth: Penguin

Tumin, MM (1953) 'Some principles of stratification: a critical analysis', *American Sociological Review*, 18, pp 387-94

Waldmann, RJ (1992) 'Income distribution and infant mortality', *Quarterly Journal of Economics*, 107, pp 1283-1302

Wennemo, I (1993) 'Infant mortality, public policy and inequality – a comparison of 18 industrialized countries 1980-85', *Sociology of Health and Illness*, 15, pp 429-46

Wilkinson, RG (1992) 'Income distribution and life expectancy', *British Medical Journal*, 304, pp 165-8

Wilkinson, RG (1986) 'Income and mortality', in RG Wilkinson (ed.) *Class and Health: Research and Longitudinal Data*, London: Tavistock

Wilkinson, RG (1990) 'Income distribution and mortality: a "natural" experiment', *Sociology of Health and Illness*, 12, pp 391-411

Wilkinson, RG (1996) *Unhealthy Societies. The Afflictions of Inequality*, London: Routledge

5 ~ Income Inequalities and Health

Commentary 1: Alison J Blaiklock

Tena koutou, tena koutou, tena koutou katoa.

I feel somewhat uncomfortable being here as a commentator. I have concerns about developing an 'industry' of people in public health, academia, health policy and health services management whose work depends on the existence of poverty and that we become commentators rather than workers for change.

One of the major issues in developing responses to poverty is participation – the involvement of people who are themselves poor in the development of solutions. This came out clearly during the Children's Coalition Conference on *The Multiple Effects of Poverty on Children and Young People: Issues and Answers* (Hassall, 1996). I cannot speak for those who with enormous courage and determination live the day-to-day reality of being poor in this wealthy country.

But we are here because we need to do something about what's happening. I'm here as a practitioner in public health working to improve the health of children and young people – and poverty is clearly a major cause of death, disease and misery for children. For example, an Australian College of Paediatrics literature review (see Jolly, 1991) found that children in the 0-4 year old age group had a range of health problems associated with low socioeconomic status including prematurity, low birth weight, poor nutrition, elevated lead levels in their blood, iron deficiency with or without anaemia, and developmental delay. The rates of stillbirth, perinatal mortality, postnatal infant mortality, sudden infant death syndrome, mortality in the one to four years age group, injury, sub-optimal growth, otitis media, fetal alcohol syndrome, respiratory infections, gastroenteritis, hospitalisation, child abuse and behavioural disorders are alarming. The children were less likely to be breast-fed, have adequate nutrition, be immunised, receive well childcare and continuity of health care, live in a safe environment, and have access to services.

Peter Saunders has pointed out some of the difficulties with information. New Zealand has no official poverty line and does not collect sufficient information to be able to participate in the Luxembourg Income

49

Study. In a study we did of children and young people in Waitakere City (Wildermoth and Blaiklock, 1996), we wanted to look at the family income for children and young people in different situations. We obtained information for different family income quintiles from the 1991 Census (Statistics New Zealand, 1995). We found that there was no information available for about one sixth of the children. This missing data limited the conclusions we could draw.

Richard Wilkinson has pointed out the problems of non-response rates in his latest book. Non-responders to questions about personal income tend to be concentrated among the poor and the rich – so that if a survey has a low response rate it probably under-represents both the poor and the rich and thus under-estimates the extent of inequality. This suggests that analysis of income distribution will underestimate the extent of inequality in countries in which there is a marked inequality of income distribution (see Wilkinson, 1996).

Peter Saunders and Ichiro Kawachi have different views on whether income inequality has an independent effect on health – that the size of the gap between rich and poor in a society affects the health status of people in that society. Peter Saunders has criticised Wilkinson's work, reworking Wilkinson's original study on cross-country comparisons using revised data from the Luxembourg Income Study. He found this removed the correlation between inequality of income distribution and life expectancy. He also described a study he has done on the relationship between income distribution and inequality in 15 OECD countries as measured by the income of the wealthiest tenth compared to the poorest tenth. This study found that, when using this measure of income inequality, there was not a correlation between income inequality and life expectancy in these countries.

Peter Saunders' criticisms demonstrate the importance of the arguments around what level of incomes to choose in making comparisons between groups – and of getting high response rates for people in those different groups. The possibility of low response rates from rich and poor people needs to be taken into account in looking at income distribution. If this could be done – and of course doing so is very difficult – then results to the studies may be different.

Ichiro Kawachi has described an important study of income inequality which makes comparisons between the different states within the United States. Two measures of income distribution were correlated with age-adjusted mortality rates. This correlation remained when adjusted for the proportion of people living below the poverty line in a state and for the proportion smoking. There was also a correlation with specific causes of death. He and his fellow researchers concluded that inequality in the distribution of income explains a significant proportion of the variation in mortality rates between states and that the size of the gap between rich and poor is related to mortality.

Both Peter Saunders and Ichiro Kawachi have proposed possible underlying mechanisms as to how poverty affects health status. Peter Saunders described an Australian study in which those living in poverty or at the margins perceived themselves to be more unhappy and in worse health than other Australians. Ichiro Kawachi and colleagues looked at the notion of social capital – the features of social organisation which facilitate people in a society co-operating with each other for mutual benefit. They found that the size of household income inequality was strongly correlated with both the per capita group membership and lack of social trust. In turn both social trust and group membership were correlated with total mortality, as well as deaths from specific causes. This study supports the theory that the mechanism by which the gap between rich and poor affects mortality is through the failure to invest in social capital.

There are a number of studies which support the hypothesis that inequality of income distribution has an independent effect on health status. There is the American study by Kaplan and others (1996) that Ichiro Kawachi referred to, which found changes over time – that the degree of inequality in a state predicted mortality ten years later. Wilkinson cites a number of studies which have linked mortality and income distribution. For example, he notes that the improvement in the life expectancy of the Japanese people between 1970 and 1989 is not adequately explained by changes in diet, smoking and GDP and may be related to narrow income differences. In England and Wales the mortality rate for infants and for adult males of working age is strongly correlated with social class, but this is much less so in Sweden. This may be because Sweden has a much smaller income distribution gap than England and Wales (see Wilkinson, 1996).

There is not yet conclusive evidence that inequality of income distribution has an independent effect on health – but the case is getting stronger. It is going to be much harder to prove conclusively than, for example, the relatively simple link between tobacco and cancer – but has potentially as much importance. As Peter Saunders has pointed out, there are major policy implications and so it is important to scrutinise the evidence carefully.

We also need to look at the overall effects of poverty and the growing gap between rich and poor on the well-being of people – not just at the effects on health. At the conference, *The Multiple Effects of Poverty on Children and Young People: Issues and Answers* (see Hassall, 1996), we looked at the effects of poverty on children's health; their education; their opportunities to develop and learn; and their ability and opportunity to participate in their family, their culture and their community.

There is strong evidence that the structure and organisation of a society impacts on the health and well-being of children and young people. Bret Williams and Arden Miller (1992) made a detailed comparison of the United States with 10 European democracies. They concluded that children's health status was better in the European countries because of their preventative health services and their social policies which provided much more support to families. A UNICEF study found there were improvements over time in the well-being of children and young people in countries using a European/Japan model of economics, whereas there was a deterioration in countries which had an Anglo-American model of economics (see Miringoff and Opdycke, 1993). This was confirmed in the UNICEF report, *Child Neglect in Rich Nations* (Hewlett, 1993).

The development of research, policy and education must involve those affected by poverty. This is both ethical and effective – those whose day-to-day lives are affected by poverty are going to know a great deal about what solutions might work. This is one of the intentions of the *Beyond Poverty* Conference[1] to be held in Auckland the weekend before the very expensive Department of Social Welfare Conference, *Beyond Dependency*.[2]

Sometimes in public health one has to act before the evidence is perfect. Many of our actions would be the same whether we deal with poverty alone or deal with poverty and inequality of income distribution. We do not need to sort out that issue before acting.

For example, a Kings Fund review of strategies for tackling inequalities in health recommended a range of actions that health services could take. These included ensuring distribution of resources between areas in proportion to relative needs (weighted capitation and taking a population approach); responding appropriately to health care needs of different groups (equity audits and reducing barriers to access); and encouraging a more strategic approach to developing healthy public policies (at a national level, inter-departmental action, intersectorial initiatives, health impact assessments of all public policies, and equity-oriented health targets; and at a local level, promoting structural links between agencies so as to affect how their policies and practices impact on health) (see Benzeval et al, 1995).

Let us remember who we are talking about – the toddler with glue ear whose parents cannot afford the prescription and whose opportunity to learn to talk is being affected, the child in an overcrowded house who is stressed out, the teenager falling asleep at school from hunger and with growing despair about their looming exams and job prospects ... . Children and young people are the group most affected by poverty in our society and they cannot wait. They have only one chance for normal development. As the churches have been saying in their poster:

For 7 out of every 10 kids
Poverty in NZ
is nothing to worry about.
For 3 out of 10
It's something to do something about.[3]

References

Benzeval, M, K Judge and M Whitehead (eds) (1995) *Tackling Inequalities in Health: An Agenda for Action*, London: Kings Fund

Hassall, IB (1966) *The Multiple Effects of Poverty on Children and Young People: Issues and Answers: A Report by Ian Hassall on the Conference held at Massey University Albany*, Auckland: Children's Coalition

Hewlett, SA (1993) *Child Neglect in Rich Nations*, New York: UNICEF

Jolly, DL (1991) *The Impact of Adversity on Child Health: Poverty and Disadvantage*, Australian College of Paediatrics

Kaplan, G et al (1996) 'Inequality in income and mortality in the United States: analysis of mortality and potential pathways', *British Medical Journal*, 312, pp 999-100

Miringoff, M and S Opdycke (1993) *The Index of Social Health*, New York: UNICEF

Statistics New Zealand (1995) *New Zealand Now: Children*, Wellington

Wildermoth, C and A Blaiklock (1966) *The Children and Young People of Waitakere City 1996*, West Auckland: Public Health Promotion

Wilkinson, RG (1996) *Unhealthy Societies: The Afflictions of Inequality*, London: Routledge

Williams, BC and AC Miller (1992) 'Preventive health care for young children: Findings from a 10-country study and directions for United States policy', *Pediatrics*, 89 (5), Supplement, pp 983-998

Notes

1 *Beyond Poverty* Conference, 14-16 March 1997, Massey University Albany.

2 *Beyond Dependency* Conference, 16-19 March 1997, Auckland.

3 New Zealand Christian Council of Social Services, Poster from Stop Poverty Now Campaign, Wellington, 1996.

Commentary 2: Brian Easton

The two papers by Saunders and Kawachi are best understood in the general context of the study of socioeconomic factors on the health status of individuals. The area in which this is best studied is that of the relationship between unemployment and sickness. In contrast to the handful of papers which Peter Saunders mentions about incomes and health, most of which were produced in the 1990s, there are hundreds of papers on the impact of unemployment, some more than half a century old (see Easton, 1990; Shirley et al, 1990).

It may seem obvious that unemployment causes sickness. Anecdote and causal empiricism report that very often the unemployed are sick and would appear to be more so than the employed. However, correlation does not prove causation.

An indication of the complications is shown in the following scheme: suppose there is an observed correlation:

- A may cause (influence) B, but
- B may cause A, or possibly
- A may cause B and B may cause A. Then again
- A and B may not influence each other, and all we have is a type II error (false positive), or perhaps
- another factor X causes both A and B.

There are all sorts of complicated possibilities, one of which is:

- X causes A and B, but Y also causes A, which also causes B. And all of this has not even tried to untangle what we mean by cause.

So it may be that unemployment causes illhealth, but it also seems likely that illhealth causes unemployment, and it may also be that some other factor causes both.

Of the numerous studies which purport to demonstrate that unemployment causes illness, each could be said to be methodologically flawed – that is, a competent social scientist could identify a number of weaknesses in the research design and interpretation which would invalidate the research claim. Nevertheless, the research offers a convincing case that unemployment causes sickness.

55

First, there are very many studies which are based on different data bases, methods, situations and countries. Each may be methodologically problematic, but collectively they overcome each other's methodological problems to offer a compelling case that unemployment is correlated with illness and that it often precedes illness.

Second, the research points to causal mechanisms to explain this correlation. The primary mechanism appears to be that the lower social status of the unemployed has psychological effects which impact on the individual's physiological and psychiatric well-being. However, interestingly – and very relevantly for these conference papers – there is not a lot of evidence one way of the other that the lower income of the unemployed is a key part of the causal mechanism.

Third, no alternative explanation which is in any way as convincing has been offered for the empirical facts.

On these standards we are a long way from demonstrating the effect of income inequality (or a more general notion of social inequality) on health status. This does not mean that there is no effect. Rather, the international research is at such an early stage that we cannot say with confidence that there is a causal process by which communities with high income inequality have lower average health status. There are few studies, each of which is methodologically problematic, which tell us much about the causal mechanisms, and it is easy to explain the observed correlations – if any – by some other means, such as poor research design, reverse causation and other factors. In summary, the income inequality research is about where the unemployment research was a couple of decades ago.

While the unemployment research is a useful component to the general area of socioeconomic inequality and health status, the hypothesis about income inequality is considerably more complicated. The research is about the employment and health status of an individual. It may be assessed at the aggregate level – as, for instance, in the longitudinal econometric studies of Brenner (1984), which are among the least convincing studies of the whole corpus. Ultimately the unemployment research is about individuals. In the case of income inequality, however, we are concerned about an individual in the context of their community. That is going to be fundamentally more difficult to research.

In addition, measurement problems abound. Before giving examples from the two papers, let me recall the huge lacuna in the unemployment literature. With very few exceptions, all the studies are about male unemployment. We probably know more about the impact of male unemployment on the health of women than we do about the impact of female unemployment on women's health. This is not a matter of explicit or implicit sexism, but arises from the problem of defining unemployment in the case of women experiencing a traditional life cycle.

Measuring income inequality is even more difficult. Non-specialists are often unaware that economists' measures are usually based upon rigorous theory. What the theory says in the case of measures of inequality is that there is no general ideal measure – it comes out of the Arrow paradox. As a result there are numerous possible ways to measure income inequality, none of which is superior to all the others. Because there is no generally ideal measure of inequality, the empirical results need either to be robust to the choice of measure, or an explanation is required to explain any divergences. Kennedy et al (1996) observes that correlations that are significant for the Pietra ratio (their 'Robin Hood' index) do not apply for a Gini index on the same income distribution. Kawachi reported at the seminar that the computations of the Gini coefficients were faulty. Otherwise the researchers would need to explain why one inequality measure gave one result, a second another. At this stage in the research paradigm, a number of different measures of inequality should be evaluated and reported.

Moreover, as Saunders' paper reminds us, the income variable we are measuring is important. Again there is a plethora of potential variables, and it would appear that Wilkinson's results are sensitive to the choice. The paper illustrates this by comparing disposable income adjusted and not adjusted for household composition. If it had explored different ways of adjusting for household composition, the results may have been even more statistically chaotic. For instance, the household equivalence scale used in New Zealand – the Jensen scale – is a priori, rather than being based upon empirical research. Its strong economies of scale almost certainly exaggerate the impact of housing costs on poverty, and underestimate the number of children who are poor.

We need to think also about the measures of health status. With the exception of the Australian material Saunders reports, the usual measure is a mortality index. Morbidity may more important, although it is harder to measure. I would urge keeping away from measures of life expectancy, which are syntheses of the experiences of different cohorts over a quinquennium or decade. I was much more impressed by Kawachi's use of age-adjusted mortality rates for specific conditions.

The problem of other variables has to be addressed. It was good to see that Kennedy et al controlled for tobacco consumption (1996). It may be that Maori who have never smoked have life expectations similar to non-Maori who have never smoked (see Easton, 1995). The results are subject to a wide margin of error, but insofar as the poor tend to smoke more than the rich, tobacco consumption may be a confounding effect on all socioeconomic inequality studies.

There is also a need to control for unemployment rates, for average income levels, for levels of real health care expenditure, for racial composition and for age composition (at the very least). It might also be worth discriminating by gender, for it may be that communities with relatively higher women's incomes may have relatively lower female morbidity.

Timing is another issue that needs attention. As Saunders reports, income inequality seems to have been increasing sharply in the 1980s while it was probably moderately decreasing in earlier decades. How are changing levels of inequality to be compounded into their effect on an instant of time? Presumably the effects of changes in inequality are not instantaneous, although unemployment appears to impact rather quickly on health.

There is a non-linearity problem here to which I refer briefly. While there seems little doubt that across all countries higher average incomes generate improved average health statuses, there are numerous exceptions. Within a group of nations with similar average incomes – say, within the rich OECD – the relationship is less evident, indicating that other factors are probably as important. Income inequality is one such other possible factor, but if it dominates income, there must be some non-linear effect going on within the income distribution. The health effects of the higher incomes of some people do not simply average against the lower incomes of others. That limits some of the possible causal mechanisms, although until these are delineated more clearly it is not evident which. As Saunders implicitly argues, the stress on the poor may be non-linear, so that above a certain level of income there is little. If so this has an important implication for poverty measurement, strengthening the sort of analysis that Townsend (1979) argues. One could also argue that health care expenditure may be more effective on the poor because they get less of it, than on the rich when diminishing returns sets in.

Even if the measures are satisfactory, or the results are robust to measurement definition, correlation does not prove causation, and it does give the mechanism which underlies the causation. Further work needs to identify those causal processes. For instance, Saunders suggests that the causal mechanism might be similar to the effect of unemployment on illhealth via status and social well-being. However, it is possible that some variable such as collective values – the sort of thing which Kawachi is trying to measure – generates policies which both reduce income inequality and provide fairer access to effective health care and other health-promoting public services. Thus there may be two quite distinct mechanisms, one through the well-being that income generates, the other through public services which correlate with lower income inequality.

I have tried to summarise what we think we know, in the following bullet-points, which contain the main effects mentioned in this review. First are a group of relationships which are not the main focus of these papers, but are a part of the whole picture:

- tobacco consumption contributes to poor health (this is one of the best understood epidemiological features);
- tobacco consumption and poverty (income or social inequality, in the context of the diagram) are correlated (but we are not sure why);
- unemployment generates poor health, probably through its impact on social well-being;
- poor health contributes to unemployment (there is a feedback loop);
- good access to health services (probably) generates good health.

The two papers have been concerned about adding the following:

- income inequality contributes to poor health, probably through its impact on social well-being;
- income inequality may contribute to poor health, insofar as it limits access to health services;
- community values contribute to good health, probably through income inequality, but also through the provision of good access to health services.

We might add a further feedback loop:

- good access to health services (if provided by the public) contributes to community values.

In summary, the whole picture is complex, and some of the critical processes are largely assumptions which at best are only moderately tested. By putting down even this simplified account we can see there is much work to do, and many statistical complications to avoid.

Like the two paper presenters, I am sympathetic to the notion that social inequality has deleterious effects on average health status. But rather than using the papers to confirm my prejudice, I have emphasised that this is a hypothesis which lacks sufficient empirical validation. These papers make a contribution but we are a long way from confirmation.

My attitude to such questions is well captured by a section in Saunders' paper which is more universal than the topic of this seminar, where he says there has been a "failure of the social policy research community to be sufficiently professional in much of its work". If the debate is served by only low quality research, it will be dominated by ideology, with ultimately any policy being determined by power and by wealth. Since the wealthy have least empathy with the poor, and the most to gain from arguments that social inequality is of no social importance, the resulting policies will be unfavourable to the poor.

Neither paper is vulnerable to Saunders' criticism, but it would be very easy to accept the papers at superficial face value in the unprofessional way that Saunders, Kawachi and I abhor.

References

Brenner, H (1984) *Estimating the Effects of Economic Change on National Health and Social Well-being: Study Prepared for the Use of the Subcommittee on Economic Goals and Intergovernmental Policy of the Joint Economic Committee, Congress of the United States.* Washington: US Government Printing Office

Easton, BH (July 11 1990) *The Epidemiology of Unemployment,* The Dean's Lecture, Wellington Medical School

Easton, BH (1995) 'Smoking in New Zealand: a census investigation', *Australian Journal of Public Health,* 19 (2), pp 125-128

Kennedy, BP, I Kawachi and D Prothrow-Stith (1996) 'Income distribution and mortality: cross-sectional ecological study of the Robin Hood Index in the United States', *British Medical Journal,* 312, pp 1004-07

Shirley, I et al (1990) *Unemployment in New Zealand,* Palmerston North: Dunmore Press

Townsend, P (1979) *Poverty in the United Kingdom,* Harmondsworth: Penguin

6 ~ An Economic Analysis of Income Inequality, Health Outcomes and the Role of Social Policy

George Barker

Measurement Issues

A number of papers at the conference surveyed and sought to advance recent national and international research on measures of socioeconomic status, deprivation and income distribution. Within this domain a particular focus of the conference was on the link between socioeconomic disparities in society and health outcomes. It is fair to say that a further narrowing of focus upon the relationship between income inequality and health occurred.

A primary question that was addressed therefore was: Does a relationship between income inequality and health exist, and what causes it or through what mechanisms does it emerge? A consensus seemed to emerge at the conference that there is currently insufficient evidence and theoretical specification of the effect of income inequality on health to support any firm conclusion. The inability to draw firm conclusions on key issues is not an uncommon state of affairs in social policy research. Indeed, a common view expressed at the conference was that there was a general failure of social policy research to be sufficiently professional in much of its work to be able to offer firm advice on key matters. I share the general concern expressed about the lack of openness, rigour, logic and objectivity in relation to social policy research.

I would suggest that underlying social policy research there should be a concern to identify ways to maximise the *well-being* of individuals, not minimise disparities *per se*. A key problem for research with this objective in mind is clearly information or measurement problems; in that there is no known cost-effective way of directly measuring well-being of individuals to inform policy choices. This suggests the need to look for proxies of individuals' well-being.

Economic theory suggests in this regard that, if we assume given prices and identical tastes, there will be a one-to-one correspondence between an individual's so-called full income and their well-being or so-called utility. Full income, or y_f, can be defined as the flow of services or returns individuals earn on the rights they may hold over tangible and intangible assets (including physical, human and social capital).

It thus includes both money income,[1] y_m, plus all forms of non-money income,[2] y_n, which people appropriate from assets they can be said to own. In short:

$$y_f = y_m + y_n$$

Full income determines an individual's *potential well-being*, by determining their consumption opportunities. Unlike well-being, it is also potentially measurable in many respects. Defined as the returns individuals reap on their rights to assets or capital, full income is also directly linked to the way rights are defined, and thus to policy. It is therefore possibly a fruitful line for any empirical inquiry, which might seek to assess the effects of policy, given the well-being objective assumed above. This is especially true given the difficulties in any alternative approach that tries to assess well-being directly.[3]

There is an important dynamic, however, which needs to be recognised, even in a full income approach. This is highlighted in the identity below which must hold for all individuals expost over time:

$$Y_{nt} \ (1+g_{nt}) = Y_{nt+1}$$

Y_{nt} = full income of individual n in time period t
g_{nt} = the growth in individual n's income between time t and t+1.

Disparities in income between individuals at one point in time (t) may therefore dissolve or be reversed between time periods because of disparities in the rate of growth of their incomes or g_{nt}. This implies that research into the relationship between income inequality and health disparities might best seek to examine lifecycle disparities in income and health to avoid measurement error. Thus, measures of income inequalities in a country at a point in time (e.g., Gini, Atkinson, Theil and Robin Hood Indices), for example, where they are based on cross-sectional data sets, tend to mask or fail to measure the degree of income mobility over time. As a result, a consistent cross-country result is that income inequality is a lot less when income is measured over the lifecycles of individuals as compared to when it is measured at a point in time.

Turning to the evidence, recent analysis of New Zealand tax and benefit data appears to confirm that there is a significant income dynamic of the kind identified here. This data unfortunately only enable one to examine the dynamics of realised market and benefit incomes of individuals, and not the dynamics associated with other forms of income relevant in a full income approach (for example, consumption of subsidised health and education services). In addition, it is drawn from two different administrative datasets, which have never been effectively combined for the purpose of policy analysis. Nevertheless, in relation to these forms of income, the following has been discovered using these data sets:

- on wage and salary income, a dynamic analysis using tax data from 1991-1994 shows that:

    i. just over 25% of those in the lowest quintile of the wage and salary distribution *at least* move on to join higher income groups in one year. This rises to just over 30% in two years (Strategic Analysis Unit, Treasury, 1995).

    ii. those with low wage and salary incomes receive on average relatively larger proportionate increases, and there is no tendency for lower (higher) than average rates of growth in income to be associated with lower (higher) rates of growth in subsequent periods; and

    iii. wage and salary income mobility in New Zealand is shown to be similar to that in other countries for whom comparable studies have been undertaken (Sweden and the UK)(Creedy, 1997).

- for those on benefits, a dynamic analysis using a sample of all those who enrolled on Unemployment Benefit, Domestic Purposes Benefit and Sickness Benefit during the June quarter of 1992, revealed that 53% left and did not return in three years.[4]

Given the obvious limits to the above information about the extent of income dynamics in New Zealand, what does overseas evidence tell us? This is worth examining to the extent that integrated data bases obtained through longitudinal household surveys are now available in a number of countries overseas and our processes of income generation do not appear significantly different from theirs at this stage.

UK evidence from a longitudinal sample of households first introduced about four years ago suggests that around 50% of those in the lowest income decile there move on to higher incomes in one year – which suggests significant dynamic movement. Indeed, over one year the group as a whole experienced a 25% increase in mean income in one year.

US longitudinal evidence enables us to take a longer view and examine not only the issue of income dynamics over a part of a sample of individuals' lifecycles, but also across generations. One US study with this focus provides the following results from a sample of adults aged 27 to 35 years old in 1988, who were children aged 6 to 15 in 1968, and who were observed for at least three years as a child and three years as an adult after age 24 years (see Corcoran and Bogess, 1997).

Table 6.1: Dynamics of Poverty in the US

	Adult Life Poor			
Child Poverty	Never	1-50%	51-100%	Total
Never	65%	9%	3%	77%
1-50%	10%	2%	2%	14%
51-100%	5%	2%	2%	9%
Total	80%	13%	7%	100%

In the above table, percentage years poor is estimated as 100 x No. of years a child (adult) is observed with family income below the poverty line, divided by the total number of years a child (adult) is observed; and family income is the sum of total family income plus the value of food stamps.

The above table highlights the following points on income dynamics:

• on adult poverty status, the bottom line suggests that 20% of the population experienced some poverty over the three-year period as adults, but only 35% of this group (or 7% of the sample) were poor for more than 50% of that period, suggesting significant movement;

• on the persistence of poverty between generations, summing across the second and third row suggests that 23% of the population experienced some poverty as a child. Summing down the left hand column across the second and third row, however, suggests that 65% of this group (or 15% of the sample) went on to experience no poverty in adult life;

- taking the incidence of the most extreme form of persistent poverty identified in the above table (or 50-100% poverty), the data suggests that only 2% of the population came from a family background that was observed poor for more than 50% of the time, and then went on to an adult life experience with the same outcome. This implies that 76% of those who grew up in families who were poor over 50% of the time went on to experience a better outcome as an adult. It is notable that this result implies not only significant upward mobility but also downward. Thus, 68% of the persistent poor as adults are estimated to have come from families who were poor less than 50% of the time, and 42% from families that were never poor.

Given the clear existence of income dynamics as highlighted in the above evidence, and its theoretical implication that disparities at an individual level may be greater or less over the lifecycle than whatever is observed at a point in time, the clear message is that the analysis of the relationships between socioeconomic inequalities, health and social policy needs to adopt a dynamic perspective and draw as much as possible on longitudinal data sets.

Before proceeding to undertake empirical analysis (and policy recommendations), however, the message from econometric theory is further clear that researchers need to clarify what theory they are relying on in terms of:

- the determinants of each individual's full income, or their financial and non-financial returns on the human, financial, physical and other capital they may own, in any period (ie., Y_{nt}); *and*
- the determinants of growth in that income (g_{nt}) over time.

If they fail to do that then they run the risk of mispecifying the relationships between variables in their empirical research and as a result reaching the wrong conclusions. This was a point alluded to by Brian Easton at the conference.

Put crudely, economic theory would tend to suggest that a dynamic exchange and investment process underlies any observed relationship between income inequalities, health outcomes and social policy as illustrated in Figure 6.1.

Fundamentally, economists tend to assume that the driving force behind any observed dynamic and any resultant outcomes (e.g., in terms of health) is the rational pursuit by individuals of their own self-interest, subject to constraints. Thus, it is the desire of individuals to maximise their well-being subject to the core problem of scarce resources that is highlighted at the top two boxes in Figure 6.1. The remainder of the figure tries to capture how the full income constraint individuals face can be relaxed over time, by their engaging in exchanges with others using the rights they may hold over various forms of capital, and through investment in the capital they hold rights to, in order to maintain and develop it over time.

The economic model then is in essence an exchange and investment model. Through economic and social exchanges highlighted in the third box from the top, individuals enter relations of mutual advantage, and simultaneously create institutions like markets, firms, families and local communities, including such entities as hospitals, schools, recreation centres, health farms and so forth. By means of these exchanges and associated institutional forms, individuals can increase their full income and as a consequence their well-being, relative to a situation of independent action.

As highlighted at the bottom of Figure 6.1, individuals can further influence their full income and therefore their well-being over time by engaging in investment activities that influence the nature and quantum of capital or assets to which they may have access over time. Human capital as identified here is one form of intangible asset which individuals may invest in, and can be defined generally to include all the capabilities embodied in individuals – both intellectual and physical – which affect their competencies. In this analysis, health is treated as a form of human capital, resulting from various investment activities. Disparities in income may then emerge in part from disparities in individuals' access to and investment in human capital, including health.

Social capital on the other hand as it is identified in the above diagram can be defined as the established set of mutually beneficial long-term relations which an individual may have, and which may be expected or relied

Figure 6.1: Economic Model for Exchange and Investment

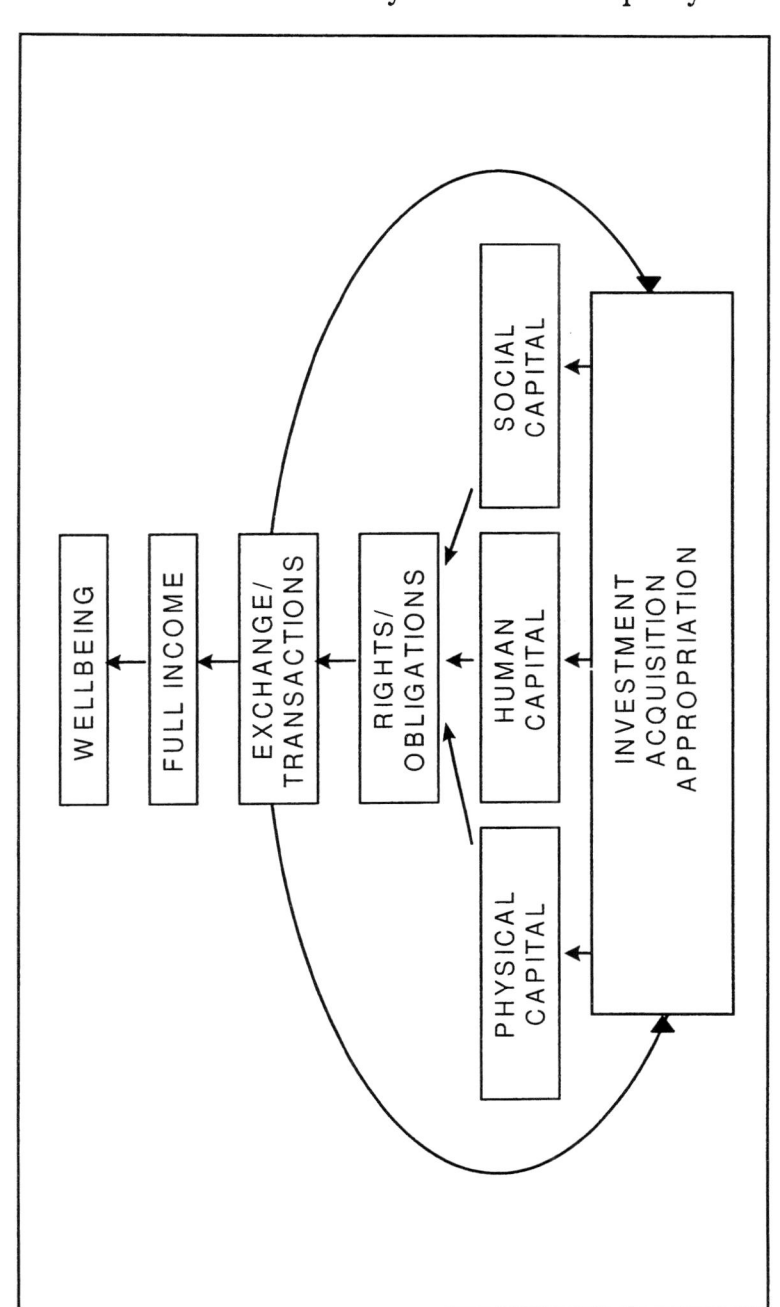

upon to support further transactions for mutual benefit. Recent economic analysis of the formation of long-term relations to support exchange recognises the existence of investment activities such as search to identify suitable parties, and subsequent relationship-specific investments to adapt relations to better support exchange over time. An economic analysis of social capital would thus see it as also arising from processes of exchange and investment subject to the rights defined by policy, as outlined in Figure 6.1. Relationship-building or the widening and deepening of relationships between individuals, may strengthen markets, firms, families and communities, and build trust, and may facilitate exchanges between individuals that enhance their income. Disparities of income in this analysis may thus emerge from disparities in individuals' access or investment in their social capital.

Figure 6.1 finally highlights the role of rights and obligations as an intervening influence on the processes involved in the generation of social and economic outcomes, including health and income disparities. This highlights the role of government policies that may influence outcomes. Thus, for example, rights to health and/or education subsidies, and obligations to pay taxes and attend school, will affect the self-interested behaviour of individuals influencing the health, employment and educational exchanges they enter into, their income in any given period, and the investments in human and social capital they make, which will in turn influence their income in later periods. Implicit in the above model, then, is the conclusion that policy will affect outcomes or the existence of disparities in income and health. This does not imply, however, that recent research which purports to show both the existence of disparities in income and health, and a relationship between these disparities, is of relevance to policy. Clarifying this point moves us on to address the second aim of the conference, namely, to discuss the relevance of findings to purchasing and policy development.

Relevance of the Research to Policy

To assess the relevance to policy of disparities in income and health and of any relationship between these disparities requires one once again to be clear on policy objectives, and to employ rigorous theoretical and empirical policy analysis. Economic analysis suggests that research on the existence and relationships between income and health disparities will have no clear relevance to policy unless it can be shown that:

i. the disparities are inefficient or sub-optimal;
ii. there are limits to private institutions;
iii. there is scope for policy improvements; and
iv. public institutions will deliver on the policies.

The following sections will address each of these points in turn, with the aim of identifying the conditions under which income and health disparities might be shown to have clear relevance to policy.

The Scope for Pareto Welfare Improvements

Economists tend to adopt the view that an economic or social outcome has no clear or easy relevance to policy decision-makers, unless it can be shown that it is possible through some change in social arrangements to make someone better off without making someone worse off. This criterion for assessing outcomes was originally proposed by the Italian philosopher, Vilfredo Pareto, and was subsequently adopted by economists. Thus, income and health disparities between individuals which cannot be changed by some means that makes one person better off and none worse off, would be seen to be efficient by economists, or optimal, and therefore of no easy relevance to policy. Economists would tend to argue that it is not disparities *per se* that are relevant to policy, but rather the scope for so-called Pareto improvements.

A key problem which would concern economists about any claimed relevance of income and health disparities to policy on efficiency gounds is why we cannot assume that any observed disparities in income and health, and relationships between these (assuming they exist), are Pareto optimal – especially given individuals' interest in their own well-being.

To answer this question would tend to lead an economist to elaborate further on the possibly hidden dimensions of the dynamic analysis outlined in Figure 6.1. These are (i) the existence of events beyond individuals' control or of risk and uncertainty; (ii) the existence of third party effects or so called externalities; and (iii) the existence of complex principal-agent relations which underlie the formation of governments, and the development and implementation of policies or laws that define the rights of individuals. The reason for doing this is that these are often identified as sources of scope for Pareto improvements in social outcomes. Figure 6.2 offers one possible perspective which seeks to capture all the above three elements.

The income generation process is again being summarised in Figure 6.2 using a two-period model. This is indicated at the base of the diagram. Any individual (quaintly drawn to the right of centre in the diagram) is assumed to travel through time, from right to left, as indicated by the continuous line running from right to left at the bottom. As indicated by lines running off this 'time line', the full income of an individual over time will be determined primarily by:

- initial holdings of tangible and intangible assets of various kinds, or 'property' rights endowments in any initial period (A_{fl});
- exchanges entered into and consumption decisions made about these that generate a monetary or non-monetary return (Y_{fl}) ;
- investment decisions implemented in the initial period (I_{fl}); and
- realised exogenous events or contingencies – either adverse or beneficial.

Figure 6.2 highlights the existence of contingencies not discussed so far, including adverse events, and therefore risk, that are likely to impact on individuals' decision-making. Individual decisions about consumption, investment exchanges which determine the values of their full assets, and their resulting full income stream over time can go wrong. In other words, events can happen to people that involve significant costs to them, including:

- loss of employment or income from employment;
- loss of health including sickness and disability; and
- loss of a partner.

A dynamic and positive analysis, however, suggests that people can take steps to mitigate these costs by:

- accumulating precautionary savings; or
- borrowing to recover from the adverse event when it occurs;
- insuring for the loss specifically with an insurer; or
- negotiating terms in collateral contracts with others to spread the costs (e.g., redundancy agreements, sick pay, marriage contracts etc.).

All the above options involve efforts by individuals to spread the costs

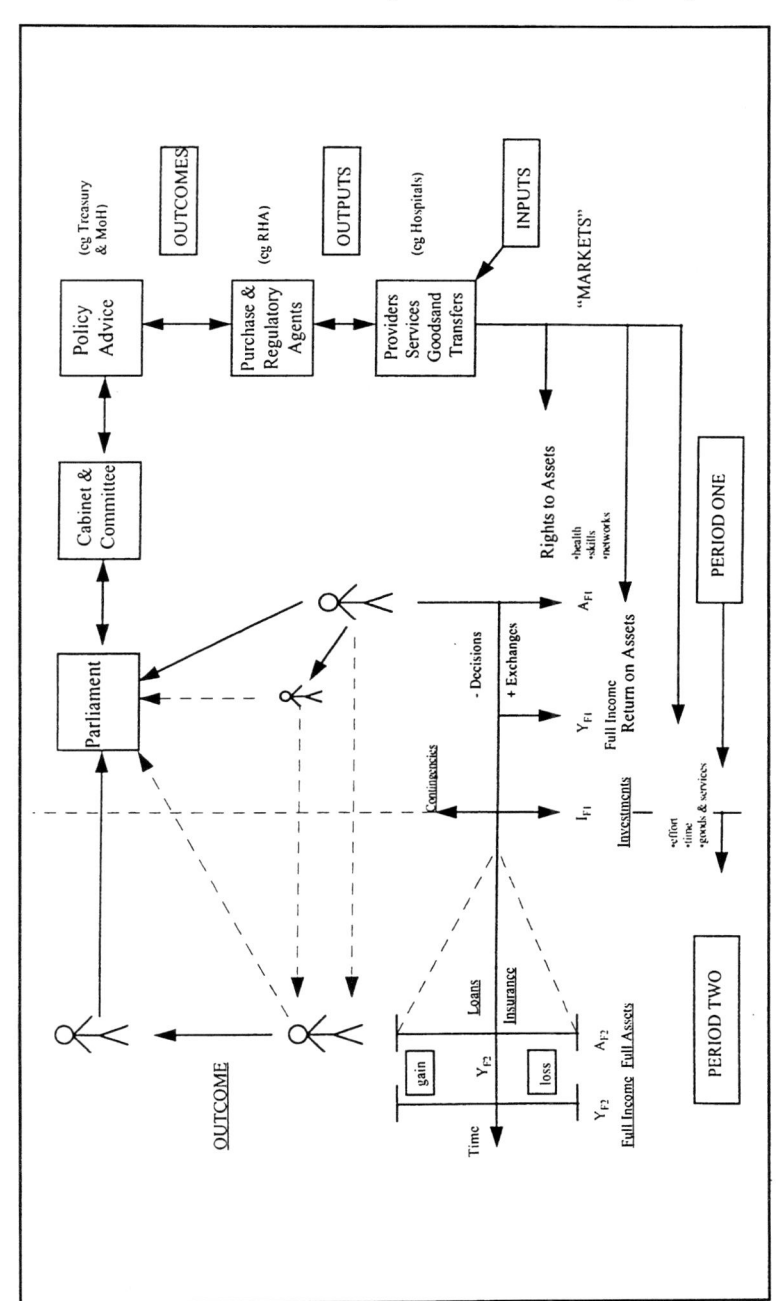

Figure 6.2: Human Capital, Principal Agent Theory and the Role of Government

of loss. The first two options are examples of attempts to spread the costs across time periods, the second of spreading the costs to others. They can all be termed equivalent transactions for dealing with loss – which, in effect, may minimise any resultant disparities.

The figure further highlights the potential for 'externalities' or for the outcomes of an individual generated by the above process not only to affect their own welfare but also those of others[5] in the second period. This is again indicated in Figure 6.2 by stick figures (this time on the left hand side of the diagram) and the arrows connecting them. As a subset of this 'externality' problem, the diagram highlights the potential impact on children's well-being in later life of decisions which parents or guardians make on their behalf, together with exogenous events. This is indicated by the smaller stick figure in the diagram. The key issue, however, is why third party effects will not be accounted for by individuals through exchanges emerging between those affected or through laws that protect the rights of third parties (including children). To the extent this occurred, any third party effects (e.g. through poor health) that arise from income disparities would be avoided or accounted for.

Finally, the potential importance of laws in defining the rights of individuals is identified at the bottom right of the diagram by arrows flowing from the right. Government policies that affect the rights of individuals also affect the consumption, exchange and investment decisions of individuals and thereby their incomes and the incomes of others. The policies generate their effect on individuals' decisions by defining and allocating initial property and human rights of individuals (including the direct use or consumption rights they hold over any tangible or intangible assets), the feasible exchanges or contracts that can be entered into in relation to these assets, the appropriability of the returns to any investments in the tangible and intangible assets, and their liability for costs (including third party costs) which their decisions may cause.

As indicated up the right hand side of Figure 6.2, in New Zealand, the policies that affect the rights of individuals and therefore outcomes, are designed by a system of representative government, which involves an elaborate set of so-called principal and agent relations between members of the electorate, Parliament, Cabinet, policy agencies, purchase and regulatory agencies, and providers of goods and services. All the way down this chain, and within every step, contracts have to be written between principals and their agents, which seek to establish mutually beneficial relations between the parties, while at the same time generating a set of policies that are optimal for society as a whole, in the most efficient way possible. This is no mean task!

Let us now return to the original question: in theory, is there any general reason why an optimal income distribution would not emerge from the above dynamic system, given that all individuals seek to maximise their own well-being? Why won't the income and health disparities which emerge be efficient, or optimal in a paretian sense, in that they could not be improved upon, without any change or gain for some being associated with some equal or greater offsetting loss in well-being for others? If one assumes that individuals seek to maximise fully their own well-being, then economic, social and political exchanges between individuals, and investment activities, should result in suboptimal or inefficient disparities in income and health being avoided. Why is it that the system will not simply settle at points where there are only irreconcilable income distribution conflicts – or the potential for only zero sum changes in income disparities, where what some gain others lose – perhaps one for one?

Although only tragic choices, or the so-called great trade-off between equity and efficiency, may have to be faced eventually – or at some theoretical point – they are less likely to confront private and public decision-makers than opportunities for improving outcomes for all over time on both equity and efficiency. This is true once one recognises that there are frictions in the social, political and economic system, much like the frictions in the physical world, that may cause divergencies from 'perpetual and perfect motion'. The term used for these frictions in economics is transaction costs. It is only the existence of transaction or exchange costs which may undermine exchange and cause problems.

Given transaction costs, income distributions which are not optimal may be expected to emerge over time, or which can be improved upon by making someone better off while not making anyone worse off, either because of health disparities or other reasons. It is transaction costs, however, and not any disparities per se, that need to become the focus of further attention.

The sources of transaction costs which may cause problems include uncertainty (or bounded rationality), the costs of information and the problem of opportunism (or incentives). These factors make social organisation towards desired goals an immensely difficult task. They have been identified in recent developments in economics as the source of significant costs in social and economic interaction. The fundamental problem, then, is to discover methods of social organisation that relax the constraints they impose in order most efficiently to marshall the activities of individuals towards common or consistent ends. To ignore the problems caused by these constraints is to ignore the fundamental issues in social organisation. It is useful, thus, to spend some time elaborating the role of each factor in turn.

If individuals could fully anticipate and understand the network of causes and effects that exist, planning, both private and public, would be a relatively simple task. The future could be anticipated and shaped, or strategies to cope with inescapable occurrences devised. Uncertainty about the future and about the consequences of certain actions derives from what may be called the bounded rationality of individuals and the inevitability of complexity and change. Bounded rationality (Simon, 1957) is the inability of humans to comprehend fully the nature of their environment, to anticipate or devise strategies to cope with change and to communicate effectively with each other. Given the existence of bounded rationality, people must plan on the basis of a largely uncertain future, and they are forced to adapt to change and adopt strategies that minimise risks. Uncertainty then becomes something to be managed, never eliminated.

Beyond the problem of bounded rationality, there is the problem of obtaining information (see Stigler, 1961; Akerlof, 1970; Alchian and Demsetz, 1972; Arrow, 1974a; Spence, 1976). People may have the ability to make correct judgements, but unless they have access to information they will simply be unable to exercise their judgement. The acquisition of information involves costs; it is not free and for this reason it is also valuable. It is therefore something to be searched out, bought and produced. As with other scarce resources, however, trade-offs have to be made. There is always some point at which it ceases to be economic to search for more information about problems or solutions. The benefits from further pursuing information need to be weighed against the costs.

A final constraint on our ability to achieve social goals is that individuals do not always have the purest or most saintly of motivations. Individuals are at least in part motivated by a concern for themselves. People thus have opportunistic tendencies (see Williamson, 1975 and 1985). The organisation of society towards the achievement of social goals would be immensely easier if individuals were always generous, altruistic, honest and forgiving. When they are selfish, jealous, deceptive and spiteful, then the costs of social organisation are raised. Individuals will therefore face difficulties in interacting and will need to protect themselves against the opportunism of others. The appropriate response to this problem is not to assume that people ought to be different; rather, it is to devise means of organisation that limit the adverse consequences of opportunistic tendencies.

The problem of opportunism can be understood as a subset or manifestation of a more general problem – the problem of ensuring that individuals face incentives that align their interests with those of others.[6] It is a problem fundamental to social organisation. How can people's incentives be aligned so that, for instance, shirking at the workplace, uncooperative behaviour, white collar crime or neglect of scarce resources is reduced?

Conclusion

Transaction cost analysis introduces greater scope for Pareto improvements than might be implied by the zero transaction analysis which characterised the nature of economic analysis which preceded it. It therefore potentially creates greater scope for research on income and health disparities to proceed to the stage of asserting the policy relevance of findings – but only to the extent that the other three conditions for proving relevance of the research findings are also shown to hold earlier. Namely:

ii. there are limits to private institutions;
iii. there is scope for policy interventions to improve outcomes; and
iv. public institutions will deliver on the policies (see p 69).

This is true to the extent that a transaction cost analysis implies the need for a so-called comparative institutional approach to policy. It requires that the role of transaction costs in generating poor outcomes are clearly shown, and that policy interventions and associated public institutions which respectively can and will improve outcomes relative to private arrangements are identified. Let us then turn to each of the remaining points determining the relevance of any research.

The Role and Limits of Private Institutions

Transaction cost analysis in economics has, at the same time as identifying the potential scope for Pareto improvements in outcomes, also been instrumental in highlighting how social but yet still essentially private (as opposed to state) institutions have developed through a process of evolution in order to minimise the problems in private exchange it has identified. This is not surprising as, in the inimical words of Arrow:

> I sometimes think that welfare economics ought to be treated as an empirical discipline. Implicitly if an opportunity for Pareto impovements exists then there will be an effort to achieve it through some social device or other. (Arrow, 1974b)

The relevant private institutions that have received attention include markets, contracts, for-profit firms, various forms of non-profit voluntary associations and the family. These institutions do not fall like manna from heaven but arise from the efforts of individuals either alone or in coalitions. In what follows, we shall discuss both the role and limits of each, and in so doing identify in broad terms how research on income and health disparities will need to proceed if it is to show how divergencies from optimal income and health disparities might still emerge, and therefore achieve relevance to policy.

Markets

In the context of simple transactions or exchanges, market contracting has many beneficial features. Markets, by using decentralised decision-making, enable efficient use to be made of information. Like traffic lights, the market price mechanism can co-ordinate individual actions and resolve interdependencies, while competition puts in place a selection mechanism that over time tends to guide resources to those users and uses that maximise the value of production secured from the resources, as measured by consumers' willingness to pay. Market competition further ensures that those forms of organisations that deliver valued outputs at least cost are most likely to survive. The resulting income distribution may tend to be efficient.

Significant resistance to the use of private markets is frequently based on the view that individuals searching for private gain will be unconstrained, resulting in undesirable consequences such as exploitation, unfair trading and monopoly practices. Such market practices, however, are often disciplined by the market itself through competition.[7] If, through exploitation or unfair trading, an individual or firm can earn a return in a particular activity that is above that earned elsewhere, then there will exist incentives for others to enter the market and compete, thereby undermining the longer term survival prospects of such practices. Thus, economic rents and privileges tend to be transient in the context of competitive processes, and are likely to develop and persist in the context of arrangements that inhibit such processes.

Private contracting nevertheless clearly faces limitations. The major sources of problems with private contracting are uncertainty, information problems and opportunism which were discussed in the earlier section. In the context of uncertainty, individuals will face problems contracting about future events. Information costs also raise the costs of contracting. In particular, individuals face costs searching out and evaluating potential contracting partners. Finally, if individuals are opportunistic or incentives are not well-aligned, then people face costs contracting with each other. For example, if one pays for goods in advance of delivery, one faces the risk that the supplier will either not deliver or deliver poorly.

These problems are unlikely to be uniformly present, however, in all exchange transactions. Voluntary contracting is thus likely to be efficient in the context of simple exchange relations such as those involved in auction markets, where there are many buyers and sellers and information is easily obtained. Capital markets exhibit these features. Problems will arise with voluntary contracting, however, when there are difficulties in measuring the quality of the goods being sold.[8] Measurement problems result from the existence of information costs and bounded rationality. They arise in goods markets where the goods are complex, but more particularly in the context of sales of services, especially expert services provided by professionals such as doctors, lawyers or accountants.

Problems also arise where the parties to the contract are engaged in a long-term relationship supported by investments specifically tailored to the trading relation and reliant upon its continuity (that is, transaction specific assets).[9] An example of such a situation includes that of a manufacturer who requires a steady supply of a unique piece of equipment not used by anyone else. In these situations there are a small number of buyers and sellers and bargaining will be complicated by the existence of information costs, bounded rationality and opportunism. The absence of large numbers of alternative buyers or sellers with a coincidence of needs means there is greater potential for haggling over the terms of trade, and for opportunistic behaviour.

Such behaviour, however, absorbs resources. Consequently, private individuals seek to anticipate such occurrences by crafting safeguards in their contracts, or by seeking out alternative organisational forms that minimise problems. Examples of such private endeavours to minimise the problems facing social and economic organisation are discussed below. The main examples are complex contracting, the firm, and the club or voluntary association.

Complex Contracting

A further form of private ordering is complex contracting. Examples of non-standard contracting practices include entry fees, marketing restrictions and some types of franchising. These practices often appear unfair at first glance and suspiciously like monopolistic restrictions and a potential source of sub-optimal income and associated health disparities. However, they may on the contrary be necessary in order to reduce uncertainty or scope for opportunism between the supplier and consumer, thereby making it economic for one party to invest in a specialised technique. Seen in this light, such practices can be efficient. Frequently, such non-standard or complex arrangements reduce the capacity of parties to a contract to renege or change the terms of an agreement. These restrictions can also be understood as efficiency enhancing, particularly when there are the measurement problems or a small numbers bargaining situation pointed to earlier. In such cases, the problems created for contractual partners by opportunism are likely to be severe. For instance, if the quality of a good or service is hard to measure, then it will be easier for one party to cheat another (see Williamson, 1985, n.5).

Reliance on a third party may be resorted to in order to deal with contracting difficulties. This is likely to take the form of privately arranged arbitration procedures that are more sensitive to the needs of the parties than the use of common law courts constrained as they are to apply general rules with little ongoing knowledge of the facts.

Perhaps the most frequent method besides arbitration relied upon by private parties to limit opportunism is the creation of what may be called credible commitments (see Williamson, op.cit.). Individual contractors may deliberately agree to commit themselves at the outset. Such commitments may serve to tie their own hands at a later date from behaving opportunistically. The arrangements likely to emerge here are similar to the mutual creation of hostages. Sometimes, however, these arrangements may appear unfair and explanations are advanced that suggest that one party is using market power to extract unfair terms. One needs to be careful to examine the detail and background to the creation of these apparently unfair terms (see Klein, 1980, n.7). They may often be better understood as attempts to safeguard integrity in a long-term relational contract. Examples include bonds required by landlords of tenants and vested pension funds offered by employers. These types of contractual arrangements can be understood as attempts by landlords to protect themselves against opportunistic tenants, and attempts by employers to protect investments they may make in the training of workers.

In addition, problems associated with uncertainty frequently give rise to complex contracting. Individuals attempt to manage uncertainty. They take out insurance contracts, they buy on futures markets, they employ specialists either full-time or on retainers, they maintain inventories or they may merely self-insure through personal savings. Clearly, attempts to anticipate all future events are costly. In designing these contracting safeguards, therefore, individuals will weigh costs against benefits, and may be expected to seek to design contractual arrangements that best suit their needs over time.

To be sure, complex contracting restrictions that are observed may simultaneously serve the efficiency purposes outlined above or other anti-social (for example, monopolistic) purposes. Here as elsewhere, however, where trade-offs are posed they need to be evaluated and the virtues of private arrangements not underestimated.

The Firm

The firm can similarly be interpreted as an organisational innovation that substitutes internal organisation of production for contracting across markets by autonomous agents, in order to minimise the costs of contracting (see Coase, 1937). Substituting administrative control for market relations may enable the pooling of information, improvements in communication and other reductions in bounded rationality, greater monitoring and control of opportunistic behaviour, co-ordination of production interdependencies and speedier resolution of contracting disputes. There may be gains both in adaptability and from improving integrity in exchange, by organising production within a firm. This is particularly likely in two cases. First, when the parties are committed to a long-term contracting relationship through the development of assets which are specific to their relationship and which have little value in alternative uses; and, second, when there are serious problems measuring the quality of a good or service provided, as in cases of team production.

With team work it may be difficult to measure the contribution of each team member, and shirking (opportunism) may become a serious problem (see Alchian and Demsetz, 1972). As the residual income or profits left after production are claimed by the owner of a capitalist firm, he or she has an incentive to monitor production and minimise production costs in order to maximise his or her income (see Fama and Jensen, 1983 a and b). Over time this operates to the benefit of the consumer as, through the increases in productivity it enables, and through competition, prices may be expected to fall.

Within the community there is often a general expression of hostility towards large dominant firms. This view is based on a suspicion that the expansion of a firm into different activities has monopoly purposes. Recent research, however, suggests that attempts to expand or take control of different levels of production and distribution may be based on the efficient adaptations of private individuals to eliminate the problems experienced with market contracting outlined above.[10]

It is true nevertheless that firms can behave opportunistically and exploit a monopoly advantage. The key question is, are there checks on this behaviour? In fact the number of firms operating in a market, or dominance, may not be the critical factor; rather, the ease with which new firms can enter the market, or the threat of competition to, or contestability of, the firm's activities may be the relevant element. If there are low barriers to entry, then checks upon monopoly behaviour will exist. Firms further need to maintain their reputation. The potential loss of reputation therefore acts as a check. Monitoring by consumers, by potential rivals, and by government (involving the threat of regulation) are all likely to act as a check on behaviour. Moreover, co-operative organisations or alternatives to the capitalist firm may be expected to survive over time if they are more productive, so long as the legal framework is neutral.

The major problem that large internal organisations create, however, is a weakening of incentives to perform. When transactions do not involve measurement problems, or specific assets, markets can be more effective. Within a large firm the checks on shirking can be comparatively low and the costs of monitoring and effectively encouraging efficient production may be weaker than when transactions are organised across markets. Further, bureaucratic rules and attitudes may hinder dynamic behaviour. These disabilities suggest that individuals may rely more on market-oriented solutions to contracting problems that may retain a higher level of competitive checks and incentives, such as those described earlier when we discussed complex contracting.

The above suggests that the firm can be understood more broadly as an organisational innovation or as a nexus of contracts between resource owners. Owners of labour, capital and land join together to provide goods or services to a market in an organisational unit that enables greater gains for each than independent or autonomous contracting.

Voluntary Associations

Other coalitions of private individuals or organisational forms can be identified which seek to pursue the mutual interests of members other than firms. These can be described as voluntary associations and include among others institutions, clubs, churches and unions. These coalitions can be understood as the creation of individuals contracting with each other to deliver a good or service of benefit to themselves as a group.[11] The services they provide may benefit their members exclusively, or benefit the wider community (for example, a club house versus a public park). While the problem of poor incentives or opportunism may arise in this context as well and limit the extent to which individuals may be able to achieve collective goals, the problems should not be over-emphasised. Indeed, in one view, non-profits are an institutional response that may facilitate altruism by donors to charities, for example. In a related view, non profits are seen to be a means of limiting opportunism by service providers, particularly in situations where consumers are poorly informed relative to sellers. Consumers in such circumstances may prefer to rely on non-profits to the extent that they may be assumed more reliable or trustworthy and less likely to exploit their informational advantage for short-term profit. Information problems in judging services of nursing homes, day care for children, blood banks, medical research, environmental protection and charities for the needy, for example, may explain the dominance in these areas of non-profits (see Weisbrod, 1988).

The main situation where opportunism is thought to undermine the effectiveness of voluntary associations arises in the case of public goods. A pure public good can be defined in theory as a good that can be used by additional consumers at no extra cost, and for which it is not possible to exclude people from consumption (for example, a radio broadcast in the absence of coding). In the case of a pure public good, poor incentives or opportunism may lead some individuals to 'free ride' on the collective benefits achieved by others from which they cannot be excluded. It is suggested that such opportunistic behaviour may lead to the instability of clubs or associations which voluntarily attempt to deliver pure public goods. This free rider problem can, however, be over-emphasised. Generally, the stringency of the conditions for a good to be truly public needs to be recognised.[12] Few goods are likely to involve zero marginal cost for use across any significant range. Further, if the individual does not participate in clubs or associations that may seek to provide public goods, then the quantity or quality of public goods that will exist will probably be reduced, hence a cost is incurred by non-participants and their 'ride' is not free but rather cheap. In the extreme, in fact, the collective good or service may not be provided at all. Moreover, whether a good is truly public or not depends on whether people can be excluded from its use. This may depend as much on the institution used to produce it as on the nature of the good. For instance, this is illustrated by decisions about whether or not to charge admission to amusement parks. Similarly, cable television has reduced the 'public' nature of television broadcasting.

The benefits of voluntary attempts to deliver goods are that they may ensure that the goods delivered are better tailored to the needs of the benefiting population. The decisions made on what goods should be provided are made by club or association members, and furthermore, if a member disagrees with the ultimate decision of the group, they can freely leave the group and seek another one that better meets their needs. In this way, voluntary coalitions may be better able to satisfy demands for goods frequently seen to be of a public nature than centrally determined solutions.

Families

The roles and limits of families are a more recent focus of transaction analysis. Increasingly, it is being recognised that families are in effect complex systems of exchange and may function as a key private institution for organising the exchange of health, education, welfare and financial assistance. Transaction cost analysis of the family as an institution and its potential role in ensuring or exacerbating income and health disparities, however, is in its infancy (see Pollak, 1985).

Conclusion

The above discussion attempts to indicate the useful role of private arrangements as means for resolving conflicts of interests and achieving collective goals even in the context of serious organisational difficulties. It attempts to point out common fallacies or misconceptions about the poor effectiveness of private arrangements. At the same time, it emphasises that private arrangements have weaknesses or face limits. In particular, private arrangements are likely to face problems arising from information costs, bounded rationality, opportunism and uncertainty. To demonstrate relevance to policy, the research on income and health disparities will need to address how private arrangements may fail over time as means for individuals addressing sub-optimal outcomes. Our discussion in the two sections which follow, however, goes on to show that the causes of transaction cost problems in private arrangements are nevertheless common sources of failure of centralised solutions. A relative assessment of the abilities of private versus centralised attempts at solutions to social problems is therefore warranted. The selective emphasis of these problems in relation to one or other institutional option should be avoided.

The Scope for Policy Improvements

We have already highlighted in general terms the potential role of policy in affecting health and income disparities, through the effect it has in determining individuals' rights to exchange and invest in assets. For income and health disparities research to be relevant to policy, however, requires that it identify specific policy interventions which may lead to Pareto improvements in the outcomes observed. It is not sufficient to show the existence of the disparities, nor the failure of private arrangements to generate fully efficient outcomes.

In this regard, one of the earliest and perhaps most fundamental achievements of transaction cost analysis was that it showed that if policies or legal rules allocating rights exist, but leave the legal rights tradeable[13] and contracting costs are zero, then it does not matter for efficiency what policy or the law actually says. In such conditions, private parties will change any rule with adverse consequences for efficiency by changing the rights created by the rule by agreement in order to exploit the gains from exchange inherent in any departure from efficient outcomes. Coase summed up this point in the so-called Coase theorem by saying that while, "... the delimitation of rights is an essential prelude to market transactions ... the ultimate result (which maximises the value of production) is independent of the legal decision" (1959, p 25).

Subsequently Cheung has taken this analysis one step further and argued that, even in a world of scarcity and interdependence, if transaction costs are zero, "... the assumption of private property rights can be dropped without in the least negating the Coase Theorem" (Cheung, 1986, p 37). This is a view which Coase commented was "no doubt right" (Coase, 1988, p 15). It also suggests that even if there is no initial allocation of rights by legal rules, efficiency will result.[14]

The Coase theorum then implies that although policy may affect income and health disparities, if transaction costs are zero and policy leaves the rights of individuals tradeable, only optimal disparities will result.[15] It thus suggests that research on income and health disparities will be irrelevant to the evaluation of policies which leave the rights of individuals tradeable, if the transaction costs of exchanging rights are low relative to the gains that may be achieved through such exchange.

This result is little more than the natural corollary of the zero transaction cost analysis presented earlier on the scope for Pareto improvements. The important additional point it highlights, however, is that prima facie research on health and income disparities may have less relevance to policies that permit trade or exchange in rights as opposed to policies or laws which limit trade and therefore are more likely to generate sub-optimal outcomes. Strangely, then, it would seem to imply that the health and income disparities researchers should tend to support highly liberal policies or legal frameworks, often associated with so-called free market ideologies, particularly if their real concern is with maximising the well-being of individuals, and not achieving equality of outcomes *per se*. This seems worth mentioning here.

As we already know, however, transaction costs are not zero and therefore the Coase theorum does not hold. Policy and law will have efficiency consequences. In what follows we shall therefore proceed to a discussion of the nature and consequences of the various means the state has at its disposal for defining allocating and enforcing rights, thereby affecting social outcomes (see Posner, 1986; Cooter and Ulen, 1988; and Mueller, 1989). To demonstrate relevance to policy, income and health disparities research will need to draw on this analysis to identify how policy may affect outcomes, and to design policy interventions that will improve outcomes.

The Courts and the Common Law

One specific policy option that should always be examined in this regard is whether the resolution of what legal rights should be created, for whom and how, should be left to court decision-making. Indeed, the courts tend to reserve the right to have the final say about disputes involving legal rights – subject to appeals to higher courts and legislative intervention – leaving this option always open to the parties to a dispute.

In comparison to relying on private ordering, what problems will judges face in settling disputes? Why should one be wary of this as a policy option – if at all? The role of bounded rationality and information problems in court decision-making is critical in this regard. Judges are human; they have difficulty forming judgements. Even if they are assumed good at it, they also face information costs. If all relevant information were public or available to all parties and the court at zero cost, then a court may be able to make a decision that approximates a value maximising exchange. Where there is information held privately by the parties, or where it is costly to access information, then such an outcome is less likely. If information is held privately by the parties, then two situations are worth distinguishing. First a situation where information is held privately by the parties, but in common. Thus both know the relevant facts, but the court does not and faces costs accessing it. The second is where the information is held privately by the parties but asymmetrically. Thus, one party holds information that the court and the other party do not

In either situation of private information, it will be difficult for a court to minimise transaction costs, where transaction costs include both the direct costs of ensuring that value-maximising exchanges of rights occur, and the costs of exchanges that may be forgone because they are too costly to consummate. The courts' capacity to perform well as an institution of dispute resolution in the instant case will depend on the substantive legal rules providing the basis of court intervention (including those of contract and tort)(see Cooter, 1985), the remedies adopted for enforcing rights (see Calabresi and Melamed, 1972), and the legal procedures applying (e.g. evidence). It will also depend on the parties' ability to bargain subsequently about the outcome to rearrange it, or appeal the decision. It would thus appear likely that poor decisions in a case will persist and have adverse effects subsequently, to the extent that information problems between the parties are likely to have caused the original decision to take the dispute to court rather than to settle privately.

One needs to add to this potential risk the likely wider impact of legal decisions on third parties, given the effect of legal precedent or the doctrine of *stare decisis*. The question then becomes whether judges will be able to settle disputes that are generally efficient, not only for the parties but also for third parties as well. Will the basis of judgements in a particular case, for example, be accurately enough stated so as to signal clearly the limits of its application, and will those limits be appropriately set? Clearly, the costs of a court accurately assessing potential third party effects will again be undermined by the information problems involved in such assessments. This will be the case even though those involved in any litigation will have incentives to identify potential adverse third party affects, basing their arguments on the public policy doctrines for legal decisions. It has been suggested, however, that the process of litigation may provide a mechanism akin to natural selection whereby inefficient legal rules are eliminated in the long run.[16] This is based on an assumption that inefficient legal rules are more likely to be overturned than efficient ones, because overturning them increases wealth, instead of potentially reducing it, or merely redistributing it.

The efficient outcome could emerge in the long-term even if judges do not consciously favour efficiency; and even if they suffer bounded rationality. It would suffice if they do not on average systematically favour either efficient or inefficient outcomes. If one assumes a neutral judiciary, two effects may then generate a bias in the common law process towards efficiency. First, it is possible to argue that rational self-interested litigants will tend to litigate inefficient decisions more frequently than efficient ones. This is true, with given costs of litigation, to the extent that the expected gains from litigating inefficient rules will be higher than efficient ones (i.e. by definition, overturning inefficient rules increases the potential value to be distributed between the parties). Second, however, it is also possible to argue not only that inefficient rules will be more frequently litigated but also that more will be spent on the litigation of inefficient rules than efficient ones. This potentially implies better prepared cases, and therefore better decisions in the particular case.

The above suggests a bias towards efficiency in the common law. This outcome depends fundamentally on the distribution of the costs and benefits of legal rules and therefore the balance of incentives amongst potential litigants. If, for example, the benefits of an inefficient rule are concentrated and the costs dispersed, then opportunism (in the form of rent-seeking behaviours) by those who stand to benefit most may lead to the litigation processes being biased in favour of the inefficient rule. Conversely, if the benefits of efficient rules are dispersed, then because the costs of litigating are concentrated, opportunism (this time in the form of free-riding behaviours by those who may benefit from a change to the rule but who wish to avoid the costs of changing it) may lead to suboptimal rates of litigation in favour of more efficient rules. Basically, no potential beneficiary of litigation may be willing to come forward to litigate for a more efficient rule.

The above suggests that the efficiency of the common law will depend on a number of factors. It may not be a proposition which is generally sustainable in all circumstances. Thus, in some situations it may be better to leave matters to private ordering, or look at other methods of social organisation. Let us then turn to consider the problems facing governments in relying on and designing other instruments of intervention. This includes taxes, subsidies, regulations and government ownership, which are all examples of instruments by which the state specifies the rights of individuals. Taxes determine an individual's right to income accruing from the use of a particular resource. Subsidies similarly affect this right, and since subsidies to one individual must be financed by taxes on other citizens (fiscal or inflationary), or by debt to be paid back by taxes on future citizens, subsidies define rights or relationships between people. Regulation here is taken to cover both Acts of Parliament and Orders in Council, which are clearly used to define rights. Government ownership similarly defines rights between individuals.

Typically, the government rarely acts through one of these alternative instruments, but through mixes of them. For example, when a government regulates, it usually becomes the owner of a regulatory agency and spends money on it. Rather than speaking of government instruments or interventions, it is perhaps more useful to speak of a policy framework or a government policy. In the following few pages, however, we briefly discuss the effects of the main policy instruments – namely expenditure, taxation, regulation and state ownership – separately. We further discuss the option of devolving decisions to local democratic control.

Taxation

The efficiency effects of taxation arise through the way they change the incentives facing individuals and firms. Where general taxation (PAYE,

GST) is used to support particular expenditures and thereby to support certain rights of some individuals (e.g. in access to education) in a general way, then the costs of raising the taxes need to be evaluated and mini-mised. General taxation used to finance such expenditure, for example, will create disincentives to work, save and invest, and encourage economi-cally wasteful activities aimed at avoiding tax (see Diewert and Lawrence, 1994). These outcomes need to be evaluated and minimised both when taxation policies are considered and when evaluating the benefits from expenditure programmes.

The great majority of taxation is of a general form designed primarily to support general government expenditure policies as above. Taxation, however, can be more specifically targeted, or used as an instrument to deter particular actions (e.g. pollution), or to protect the rights of others. In this case it is important to evaluate not only the effects of the taxation on the targeted behaviour but also what is done with the revenue, and the economic effects of the use made of the expenditure on incentives and behaviours of other agents.

Expenditure

The efficiency costs of expenditure can be direct or indirect. The direct efficiency costs of a subsidy arise through the way it alters incentives for individuals. Subsidies may discourage work, or encourage investment in areas where it would not otherwise have been undertaken. There will thus typically be a forgone activity or opportunity cost that needs to be incor-porated into decision-making that involves government expenditure. This will critically depend on the method used to deliver expenditure. Thus, the terms on which government money is made available to individuals (e.g. beneficiaries), or the terms on which the government's purchases of outputs from firms are negotiated or regulated will critically affect incen-tives and outcomes.

Expenditure, moreover, needs to be financed. The indirect costs of expenditure will thus also depend on the way it is financed. These costs will have to be borne by economic actors, and essentially they illustrate the interdependent nature of the economy. The indirect efficiency costs of financing expenditure when taxation is used were discussed in the last subsection. Alternative financing mechanisms include debt financing and the creation of money.

The indirect efficiency costs of debt financing relate to the effect that public sector borrowing has on the supply of capital and rate of savings in the economy. These are affected through various mechanisms, but primarily through interest rates and expectations. Large public sector debt programmes put pressure on interest rates that adversely crowd out private sector activity. Debt financing will also place a burden on future generations and therefore have intergenerational equity effects (see Auerbach et al, 1995).

If expenditure is financed by creating money, it will cause inflationary pressures involving efficiency and equity effects. If the inflationary effects of expanding money supply to finance expenditure are not anticipated by economic agents (e.g. built into interest rates), then it may create advantages for borrowers, and disadvantages for savers and lenders – with potentially adverse effects for borrowing and saving decisions. The inflation created by this means of funding expenditure will also tend to increase the uncertainty facing firms making investment decisions. This is true to the extent that they will have difficulty judging whether price rises are general or relate only to their own markets.

Regulation

Regulation[17] by central government also can be used to define the rights and obligations of individuals Provision then needs to be made for the rights and obligations to be administered or enforced, which may occur by administrative agencies and/or the courts. Regulations effectively impose a quasi-tax or cost on some, and confer a quasi-subsidy on others. They can then have adverse efficiency effects if they allocate rights poorly, or reduce exclusivity of property rights; if they prevent more effective private arrangements emerging by raising the costs of contracting, or create barriers to mutually beneficial trades where they are appropriate; and if they fail to provide for the successful administration or enforcement of rights. They can also potentially have equity effects if they serve to protect privilege, or prevent access to opportunities for some groups.

State Ownership

A further instrument for government policy involves the assumption of ownership rights over agencies or firms. This gives government residual property rights over the assets of the firm, and therefore rights to control directly decisions relating to the income, use and transfer of the assets. As indicated in the earlier section on private arrangements, ownership may be appropriate where there are significant costs to contracting between autonomous agents. An owner of a firm or organisation may typically have access to better information and achieve greater adaptability in securing supply. However, while ownership can be beneficial it can also be costly. Fundamentally the incentives of managers and employees of organisations are difficult to align with the goals (of the owners) of the organisation. This arises because of the problem of opportunism. Individual members of organisations have a tendency to pursue their own goals, to shirk and to featherbed, and to pay insufficient care in the use of resources that are owned by 'the organisation' or someone else. This category of problems with ownership can be classed as principal-agent problems.

Whereas firms engaged in competitive or contestable markets supplying goods to a consumer face checks on their behaviour and need ultimately to serve consumers as well as other firms can, a state-owned enterprise even in a contestable market can have greater leeway. This results ultimately from the fact that the state as owner has a tendency to underwrite losses and a greater capacity to do so through its ability to tax. While some checks involved in the managerial market or the ability to hire and fire managers are present for both state and privately owned organisations, state-owned organisations lack some of the controls present in most private firms and are frequently given conflicting objectives.[18] Large private firms on the other hand are typically open corporations, with their shares being traded on the capital market. If managers in a company do not maximise the goals of owners, the company's share price will tend to fall. The value of the capital assets, or the cash flow of the firm in alternative uses, will at some point eventually exceed the share price and a takeover bid is likely to follow. The first change a successful takeover company is likely to make is to restructure the organisation and fire existing managers. This capital market check then serves to keep managers of private firms from pursuing sub goals and provides a mechanism through which control and monitoring systems are improved. These checks do not exist in the state sector to the same extent.

Ownership therefore has costs, and the point to be borne in mind is that frequently social objectives may be more efficiently achieved through subsidies or taxes, or regulation of privately owned companies or contracts with them, rather than through state ownership. In those areas where the state cannot adequately secure an objective without ownership or there are net gains to ownership, then the state needs to be concerned on an ongoing basis to review and improve its internal management control systems.

Local Democratic Control

Local democratic decision-making is frequently proposed as a possible alternative means for resolving conflicting interests and achieving collective goals, while avoiding many of the problems of extensive central government involvement. When local democratically elected units result from private arrangements, the relevant perspectives for assessing such outcomes are those discussed earlier in relation to voluntary associations. Devolution of decision-making power by central government to locally elected bodies through a centrally determined legislative framework, however, is the focus of our attention here. Although in effect it represents a case where the various instruments described above are being used in unison, the devolution of decision control to a local democratic unit warrants its treatment as effectively a separate policy instrument of a unitary central state such as that which exists in New Zealand.

The main advantage offered for decentralising control to local bodies is the likelihood that better solutions will result if those making decisions are close to the problems being addressed. This is based on the notion that those who have the information should make the decisions. The existence of different jurisdictions offering different mixes of output, quality and cost then enables voters to vote with their feet (see Tiebout, 1956), as well as through the ballot box to ensure accountability of decision-making. This argument is used in support of all forms of decentralisation.

There are a number of reasons, however, for concern with the effectiveness of this particular institutional option. Fundamentally there is a need to be sensitive to the method by which decentralisation is achieved in these matters (see Oates, 1972). In many cases, the local electorate typically appears to exhibit weak interest in the prudent use of resources by elected managers, particularly in cases where central government provides the bulk of resources which the local body administers. It is possible to suggest that so long as either or both the expenditure or control exercised by the local body is laid down from the centre, the local electorate is likely to be relatively divorced from or disinterested in the management process. So long as the finance for expenditure is not raised locally by the local body, and so long as the source of the body's ultimate control does not lie with the body itself but with the centre, the local electorate is likely to exhibit less interest in holding the locally elected members to account. A kind of rational disinterest may be expected to prevail.

Public Institutions that Can Deliver

So far we have suggested that for income and health disparities research to be relevant to policy, it needs to demonstrate both the limits to private arrangements, and the scope for specific policy options to improve outcomes. We have further identified in general terms how the problems of scarcity, interdependence, bounded rationality, information costs and poor incentives can undermine the effectiveness of private arrangements, and create scope for improvements in the design of specific policy interventions.

Exactly the same problems that compound or undermine private arrangements and create scope for improvements in the design of specific policy interventions, however, can undermine our ability to rely on central government attempts to improve the outcomes of private arrangements, or supplant private endeavours. In many ways, however, the problems may be more severe. Let us then turn to consider the problems facing central government decision-making because of bounded rationality, information costs and incentive problems..

To set the stage for analysing the sources of failure in government institutions, let us begin with a government based on direct democracy (no representative government). If we assume a world of direct voting on a single issue under simple majority rule,[19] under these conditions it can be shown that the median voters preferences will be decisive in determining the collective outcome.[20] But there is nothing in the voting process to ensure the median voter's decision is the efficient one. The decision about the desired intervention depends on the individual's perceived benefits from the intervention (e.g. expenditure or regulation) and her expected share of any cost (e.g. in taxes or forgone output) for a given level of intervention.[21] For the median voter's choice to be efficient, however, the summed benefits across all voters must be equal to the marginal cost of the intervention. Further, each individual's cost shares must be adjusted so the cost price is equal to the benefits she perceives from the level of intervention chosen by the median voter.

The sources of inefficiencies arising from government are even more complex when we go to a system of representative democracy. Representative democracy is the chosen method of government in most western democracies. It is therefore the basis of the discussion of government that follows. It has the clear advantage of allowing specialisation, but because accountability is muted,[22] it also involves costs.

Like private individuals, decision-makers in a representative government will face difficulties dealing with the problems of scarcity and interdependencies that underly problems of social organisation. The reason, of course, is that the state is made up of individuals subject to the same limitations as private economic actors.

Indeed, the problem arising from the *bounded rationality* of individuals may be more significant when one considers central decision-making or planning. Even assuming central planners are as rational or more rational than others in the economy, a number of problems are likely to undermine the efficacy of central decision-making. Given the typically economy-wide effects of government actions, and the complexity of the problem solving state decision-makers are expected to engage in, more severe demands may be placed on their bounded rationality than is the case for private planning.

Centralised decision-making will also face major *information* disabilities. The information relevant to a central planning decision may be hard to obtain. The information relevant will typically be diverse, and may include unavailable information on consumer preferences, or alternative production technologies, or alternative ways of organising activities. Alternatively information may be possessed by individuals who are difficult to locate, or be of a nature that is difficult to communicate from one agent to another. If information is difficult to transfer then this means that it will be difficult both to acquire information at the centre relevant to decisions to be made, and then disseminate that information from the centre to the agents who are to carry out plans. The information costs underlying centralised decision-making therefore militate against its successful execution. It is likely to be based on incomplete information with consequent adverse effects.

In a decentralised setting, individuals accept risks, make judgements and adapt to unexpected occurrences, with those who are caught out suffering the consequences of their mistakes, and those who adopt successful strategies benefiting. There are thus strong incentives for individuals to seek out successful strategies, with successful strategies being likely to survive, thereby benefiting the system. In comparison with central planning, mistakes tend to be costly and impact on everyone, with few alternatives being available when things go wrong. The only safeguard is conscious and purposive policy review. *Incentives* to undertake this effectively, however, in order to conserve on scarce resources may generally be weak within the state, given the fact that state decision-makers may not bear all the costs of poor decisions. The state's ability to tax may mean its decision-makers do not really face the true cost of resources with consequences for its use of resources and for other sectors in the economy. Similarly, when it regulates or owns resources or organisations, its incentives to monitor the effects of these policies on the use of scarce resources efficiently may be weak.

The incentive problems undermining decision-making in 'the state' in relation to property rights go deeper than this, however. We have already noted that *opportunism* creates incentive problems for private arrangements. Opportunism in the context of central government decision-making and planning may exhibit itself in the form of political favours, featherbedding and waste of public resources. These problems are fundamental to an evaluation of the most suitable level and form of centralised state decision-making in a society. The mechanisms through which opportunism and incentive problems generally work in the state are in fact more subtle and complex than is widely appreciated.

On the one hand, definitions of property rights impact on the overall growth of the economy; on the other hand, they impact on the rents accruing to particular groups. If coalitions of individuals can either gain direct control of the property rights defining power of the state or bargain with the state to change property rights (for example, Federated Farmers, Business Roundtable, Federation of Labour, Public Service Association), then they can affect the value of their property rights, or acquire better rights (through tariffs, taxation, subsidies or legislation).

From the redistributive societies of ancient Egyptian dynasties, through Greek and Roman times to the medieval manor, there was a persistent tension between the ownership structure that maximised the rents to the ruler or particular groups, and an efficient and equitable system that encouraged economic growth and social equity (see North, 1981). The dominance of agriculture in the western world prior to the nineteenth century resulted in struggles to control the state being associated with the distribution of landed wealth and income. Over more recent times, changes in the dominant interests within society and the growth of pluralism have been associated with the radical changes in relative prices stemming from the Industrial Revolution. These changes have generated different outcomes in terms of property rights conflicts. The decline in the relative importance of land rent (and the landlord), the growth of manufacturing and services, and the growing share of income going to labour, transformed the structure of production and created new interest groups.

Property rights conflicts may appear more subtle now that they are largely not associated with tangible resources like land, and also more complex given a pluralistic society. The nature of the interventions affecting the value and distribution of property rights are also far more developed including taxation, government expenditure, regulation and legislation. However, conflicts of interest and the state's role in resolving them are no less critical to the efficiency and equity of our economy than earlier ones. The state is a double-edged sword. It can pursue generally accepted social goals or it can be diverted to pursuing the interests of particular groups.

In general, government policy faces the danger of two types of capture: from external sources – that is, lobby groups;[23] and from internal sources – that is, its own bureaucracy (see Niskanen, 1971). The mechanisms of policy capture and safeguards against it need to be continually reviewed.

One of the clear ways through which inappropriate policy may result from external capture lies in the differential effects of policies on people. Frequently benefits are concentrated on particular groups while the costs of the policy may be dispersed. This sets up a dynamic process where those who benefit from the policy find it easier to organise and lobby for its introduction and maintenance, while those who bear the costs of the policy find it too difficult to organise an effective opposition.

The relation between a government and its own bureaucracy on the other hand can be described as a bilateral monopoly. It can be suggested that the relation favours the bureaucracy, that elected representatives are at a disadvantage in relation to their own bureaucracy simply because of an information asymmetry (see Niskanen, 1971, n. 45). The problem that the bureaucracy may hold better information about how government services actually operate creates the potential for opportunism or subgoal pursuit by the bureaucracy, including shirking, budget maximisation and generally inefficient policies for society as a whole.

Thus, the main factors that will influence the extent, nature and outcome of central government intervention in New Zealand are:

i. the nature of representative government adopted, including the powers of the executive and its relationship to Parliament, and questions relating among other things to a Bill of Rights and the electoral system; and

ii. the structure, organisation and accountability of the civil service.

Propositions about what services the government ought to finance are reduced in relevance in the absence of an analysis of how the above system actually performs. Similarly, however, questions about how the government should deliver whatever services are deemed necessary require an assessment of the way government delivery actually works, what are the pressures exerted upon it – what are the constraints, the relationships between action and outcome, and the preferences underlying government activities. For example, what are the consequences of maximising behaviour on the part of economic agents within the government's bureaucracies?

A comparative systems analysis on government itself would distinguish the effect of alternative institutional arrangements on incentives and information and thus on behaviour at two levels:

i. the political institutions which express the preferences for public services;

ii. the bureaucratic or other state organisations which supply these services.

The integration of these two elements, or the process of exchange between these institutions, how it is structured and conducted, is a further focus of attention. The key element here is the nature of the relation and the implications of this for management systems.

Conclusion

The analysis of government and government policy needs to be based on a comparative systems approach. This approach invites assessing alternative institutional structures (both private and governmental) according to the processes and outcomes they involve, utilising generally accepted criteria for making social choices. This will require in-depth consideration of the goals of our society and of the means to achieve them. In comparing different means or institutions one should assess primarily:

A   their efficiency implications by examining in particular:

   i.   the incentives they create;
   ii. their effect on the efficient use of information;
   iii. the evolutionary or dynamic adaptability characteristics of the institution; and

B   their equity implications by examining:

   i.   their effect on the opportunities of individuals;
   ii. their effect on the fairness of outcomes;
   iii. their effect on the fairness of processes.

A comparative systems approach is 'level headed' about the limits of government and the limits of private arrangements; eschews the blind pursuit of ideal worlds by recognising trade-offs between goals; and places emphasis on a detailed microanalytic approach or, simply, attention to detail including empirical evidence and argument. The comparative systems approach to policy formation, moreover, suggests that there may be little one can say about the appropriate 'level' of government involvement in the economy, whether it should be more or less, without conducting a comparative systems analysis in every area of policy.

The comparative approach to formulating policy then involves two steps. Firstly, identify the objectives and rationale of government policy with an emphasis on identifying clearly the nature of the problems it seeks to address. Any claimed problem with private arrangements should be subjected to detailed scrutiny. In the past the limits of private contracting have often been overstated. The need for empirical research, or information and analysis of private arrangements that may deal with perceived problems, is central. It should not be assumed, for example, that particular groups are not competent to look after their own interests without some convincing evidence that extends beyond anecdotal observation. Further, every perceived problem should be subjected to detailed analysis as to its causes. For instance, claims that there are divergences between social and private costs are not useful ways to express many perceived problems.

Secondly, it is not sufficient to establish that a problem exists; it is also essential to establish that government can improve things. The essential element in policy formulation is therefore a detailed comparative systems approach. This comparative analysis needs to be based on an evaluation of the full impact of a government policy over time and its consequent impact on other areas of the economy. This suggests the need to avoid partial analysis of policy interventions and to adopt a more critical attitude to the likely hidden or secondary effects of any policy, given the problems of interdependencies repeatedly pointed to in this chapter.

In the design of policy, a comparative systems approach moreover suggests that attention should focus on minimising the problem of 'government failure'. Tendencies for governments to fail in the achievement of generally accepted objectives and for them to pursue other objectives can be reduced, although not eliminated, by institutional design based on a comparative systems analysis of the way the government organises itself.

In undertaking this analysis, the basic principles to keep in mind are:

(a) *Clarify objectives:* It is important that the objectives underlying any particular area of government activity are clear.

(b) *Transparency:* There is a need to ensure that there is transparency not only in the objectives being pursued but also in the means by which objectives are to be achieved. This implies, for instance, the need to make explicit possibly hidden subsidies.

(c) *Avoidance of capture:* There is a need to minimise scope for the capture of government policy when designing both the structure and processes used to formulate and deliver government policy.

(d) *Incentives:* There is a need to ensure incentives on individuals in the state are aligned to the achievement of government goals.

(e) *Information:* It is important that efficient use is made of information, and that the costs of information are adequately recognised.

(f) *Accountability:* The design of incentive and information systems should attempt to enhance accountability of the government's agents to their principals, namely Ministers and ultimately the electorate.

(g) *Contestability:* Where possible, in order to enhance both incentives and the efficient use of information, contestability of both policy advice and service delivery should be encouraged, either externally or internally.

Noble and clear objectives are not all that is needed for a government to perform well. It also requires a clear understanding of the nature and effects of its policy instruments and of the potential and the limitations of both private arrangements, and central control.

References

Ackerman, SR (ed.) (1986) *The Non-profit Sector:Economic Theory and Public Policy,* Oxford: Oxford University Press

Akerlof, G A (1970) 'The market for lemons: quantity, uncertainty and the market mechanism', *Quarterly Journal of Economics* , 84, pp 488-500

Alchian, A and H Demsetz (1972) 'Production, information costs, and economic organisation', *American Economic Review,* 62 (December), pp 777-95

Alchian, A and S Woodward, (1987),'Reflections on the theory of the firm', *Journal of Institutional and Theoretical Economics,* 143 (1), pp 110-36

Anderson, TL and PJ Hill (1975) 'The evolution of property rights: a study of the American West', *Journal of Law and Economics,* 18, pp 163-79

Arrow, K (1974a) 'Limited knowledge and economic analysis', *American Economic Review,* 64, pp 1-10

Arrow, K (1974b), *The Limits of Organisation,* New York: WW Norton

Auerbach, AJ et al (1995) *Generational Accounting in New Zealand: Is there Generational Balance?,* Paper presented to New Zealand Economists Association Conference, Lincoln University, August 1995

Barzel, Y (1985) 'Transaction costs: are they just costs?', *Journal of Institutional and Theoretical Economics,* 141 (1), pp 4-16

Becker, GS (1983) 'A theory of competition among pressure groups for political influence', *Quarterly Journal of Economics,* 98 (August), pp 371-400

Black, Duncan (1948) 'On the rationale of group decision-making', *Journal of Political Economy,* 56, pp 133-46

Buchanan, JM and G Tullock (1962) *The Calculus of Consent,* Ann Arbor: University of Michigan Press

Buchanan, JM (1965) 'An economic theory of clubs', *Economica* , 32, pp 1-14

Calabresi G and AD Melamed (1972) 'Property rules, liability rules and inalienability: one view from the cathedral', *Harvard Law Review,* 85, pp 1089-1128

Cheung, S (1986) 'Will China go 'capitalist'?, Hobart Paper 94, London: London Institute of Economic Affairs

Coase, R H (1937) 'The nature of the firm', *Economica,* 4 (November), pp 368-405

Coase, RH (1959) 'The Federal Communications Commission', *Journal of Law and Economics*, 2, pp 1-40

Coase, R H (1974) 'The lighthouse in economics', *Journal of Law and Economics*, 17 (2) (October), pp 357-76

Coase, R H (1988) *The Firm, the Market and the Law*, Chicago: University of Chicago Press

Cooter, R (1982) 'The cost of Coase', *Journal of Legal Studies*, 11 (January), pp 1-34

Cooter, R (1985) 'Unity in tort contract and property: the model of precaution', *Californian Law Review*, 73 (1), pp 1-51

Cooter, R and L Kornhauser (1980) 'Can litigation improve the law without the help of judges?', *Journal of Legal Studies*, 19 (1), pp139-63

Cooter, R and R Ulen (1988) *Law and Economics*, Illinois: Scott Foresman & Co

Corcoran, M and S Bogess (1997) 'The intergenerational transmission of poverty and inequality; a review of the literature', in George Barker (ed.) *Cycles of Disadvantage*, Wellington: Institute of Policy Studies

Cornes, R C and T Sandler (1986) *The Theory of Externalities: Public Goods and Club Goods*, Cambridge: Cambridge University Press

Creedy, J (1997) *Statics and Dynamics of Income Distribution in New Zealand*, Wellington, Institute of Policy Studies

Diewert, WE and DA Lawrence (1994) *The Marginal Costs of Taxation in New Zealand*, Canberra: Swan Consultants Pty Ltd

Downs, A (1957) *An Economic Theory of Democracy*, New York: Harper and Row

Fama, EF and MC Jensen (1983a) 'Separation of ownership and control', *Journal of Law and Economics*, 26 (June), pp 301-26

Fama, EF and MC Jensen (1983b) 'Agency problems and residual claims', *Journal of Law and Economics*, 26 (June), pp 327-51

Klein, B (1980) 'The transaction cost determinants of "unfair" contractual arrangements', *American Economic Review*, 70 (May), pp 356-62

Mueller, DC (1989), *Public Choice II*, Cambridge: Cambridge University Press

Niskanen, W (1971) *Bureaucracy and Representative Government*, Chicago: Aldine

North, DC (1981) *Structure and Change in Economic History*, New York: WW Norton

Oates, WE (1972) *Fiscal Federalism*, London: Harcourt Brace

Olson, M (1965) *The Logic of Collective Action,* Cambridge: Harvard University Press

Pollak, R (1985) 'A transaction costs approach to families and households', *Journal of Economic Literature,* 3, pp 581-608

Posner, R (1986) *The Economic Analysis of Law,* 3rd edn, Boston: Little Brown & Co.

Pratt, JW and RJ Zeckhauser (eds) (1985) *Principles and Agents: The Structure of Business,* Boston: Harvard University Press

Priest, G L (1977) 'The common law process and the selection of efficient rules', *Journal of Legal Studies,* 6 (January), pp 65-82

Priest, GL (1980) 'Selective characteristics of litigation', *Journal of Legal Studies,* 9 (March), pp 399-421

Priest, GL and B Klein (1984) 'The selection of dispute for litigation', *Journal of Legal Studies,* 23 (January), pp 1-56

Ross, S (1973) 'The economic theory of agency: the principals problem', *American Economic Review,* 63 (2), pp 134-9

Rubin, PH (1977) 'Why is the common law efficient?', *Journal of Legal Studies,* 6, 51-64

Shavell, S (1982a) 'Suit settlement and trial: a theoretical analysis under alternative methods for the allocation of legal costs', *Journal of Legal Studies,* 11 (January), pp 55-81

Shavell, S (1982b) 'The social versus the private incentive to bring suit in a costly legal system', *Journal of Legal Studies,* 11 (June), pp 333-39

Simon, HA (1957) *Models of Man,* New York: John Wiley & Sons

Spence, AM (1976) 'Job market signalling', *Quarterly Journal of Economics,* 87, pp 355-74

Stigler, GJ (1961) 'The economics of information', *Journal of Political Economy,* 69 (June), pp 213-225

Stigler, GJ (1971) 'The economic theory of regulation', *Bell Journal of Economics and Management Science,* 2, pp 3-21

Teece, DJ (1982) 'Towards an economic theory of the multiproduct firm', *Journal of Economic Behaviour and Organisation,* 3 (March), pp 39-64

Tiebout, CM (1956) 'A pure theory of local expenditures', *Journal of Political Economy,* 64, pp 416-24

Umbeck, J (1981) 'Might makes rights: a theory of the formation and initial distribution of property rights', *Economic Inquiry,* 20 (January), pp 38-59

Weisbrod, BA (1988) *The Non-profit Economy,* Harvard University Press

Williamson, OE (1975) *Markets and Hierarchies: Analysis and Antitrust Implications,* New York: Free Press

Williamson, OE (1985) *The Economic Institutions of Capitalism,* New York: Free Press

Notes

1   *Money income* (y_m) comprises the monetary returns an individual earns on their property rights to their own human capital (including wage and salary income), plus the monetary returns they earn on their property rights to financial assets (e.g., dividends and interest).

2   *Non-money* income (y_n) includes the flow of services from the rights they hold over physical wealth (e.g., imputed rents on owner occupied homes); the value they derive from using their property rights in own production; the more intangible and possibly collateral returns they may reap from the exchange or use of their rights (including job satisfaction from the particular employment contract they enter); and importantly their enjoyment of their rights to control their own leisure time.

3   The problem that must be carried forward with this approach, however, is the one already noted. Basically, information on the nature of a large number of the assets over which individuals hold valuable rights, and which affect their full income, may not be observable at reasonable cost. Such relevant information would also tend to be manipulated by individuals, especially if their entitlements were to become based on them.

4   Work undertaken by the Strategic Analysis Unit, Treasury, and The Social Policy Agency.

5   This could be through simple externalities felt by others and generated by the behaviours of the individual given the institutional framework or system of rights in place (e.g. the incidence of crime or communicable diseases or tax liability effects borne by third parties). Or because of interdependencies due to altruistic or less virtuous preferences held by others about the well-being of the individual.

6   A key early part of the literature focusing on this problem was principal agent theory. The term principal-agent problem is due to Ross, 1973; for an early review, see Pratt and Zeckhauser, 1985.

7   For a discussion of 'unfair' contractual terms, see Klein, 1980.

8   For a discussion of the problems information and measurement costs create for the operation of markets, see Akerlof, 1970; Barzel, 1985.

9   For an analysis of contracting problems in the context of specific assets, see Williamson, 1985, and Alchian and Woodward, 1987.

10 For a transaction cost explanation for multiproduct firms, see Teece, 1982.

11 Two seminal articles on voluntary associations are Buchanan, 1965, and Olson, 1965. For a review of the literature on clubs, see Cornes and Sandler, 1986. For recent work analysing the role and economics of non-profits, see Weisbrod, 1988, and Ackerman, 1986.

12 For a sceptical view on the evidence on the existence of public goods, see Coase, 1974.

13 Whenever legal rules may be waived by agreement, or can be amended by private parties, then the legal rights they establish can be termed 'tradeable' or alienable.

14 For empirical work on the private formation of property rights, without the state, in a positive transaction cost world, see Umbeck, 1981, and Anderson and Hill, 1975.

15 See Cooter, 1982, for a discussion of 'Hobbesian' versus 'Coasean' bargaining assumptions that may characterise differences in view on this.

16 For seminal articles on this see, Rubin, 1977, and Priest, 1977. See also Priest, 1980; Priest and Klein, 1984; Cooter and Kornhauser, 1980; and Shavell, 1982a and b.

17 The term is used here in the broad sense as adopted by economists. For a seminal article on the economic analysis of regulation, see Stigler, 1971.

18 As a result, not only is there a lack of incentives or rewards for monitoring, but also monitoring is more difficult or costly.

19 For a discussion of the costs of decision-making under alternative collective decision rules to majority rule, see Buchanan and Tullock, 1962.

20 For simplicity we are ignoring the problems of stability of majority rule outcomes. Majority rule can result in what is known as cyclical outcomes – outcomes that vary depending on the ordering of presentation of choices – when voters preferences are not single peaked. See Black, 1948, and Downs, 1957.

21 Another source of inefficiency arises when the voting population does not coincide with the tax-paying population.

22 To the extent that voters, for example, have to vote on a representative stance on a package of issues.

23 For a formal model of interest group lobbying and references, see Becker, 1983.

7 ~ Measuring Poverty in New Zealand

Robert Stephens and Charles Waldegrave

Introduction

Recent health reports indicate an increased incidence of poverty-related diseases such as tuberculosis, infant mortality, rheumatic fever, meningoccal disease, asthma, glue ear and iron-deficient anaemia. There are claims that the stress associated with poverty has also been a contributory factor to the greater number of reported cases of domestic violence, child abuse, home-alone children, marriage break-ups, sole parentage, youth suicide and child crime. Each of these factors will have a potential impact upon health status.

To determine whether these links between poverty and illhealth are significant requires clinical knowledge of the causes of diseases, psychological analysis of reasons for dysfunctional families and an independent measure of poverty. This paper considers the latter topic, and reports on the approach taken by the New Zealand Poverty Measurement Project to develop a poverty measure (Stephens, Waldegrave and Frater, 1995; Waldegrave, Stuart and Stephens, 1996). They established a poverty measure based on the views of contemporary New Zealand society, and related to current economic and social conditions and policy parameters in New Zealand. This discussion on the use of focus groups to draw upon their collective experience as to the income required to achieve a minimum adequate household expenditure follows a brief discussion of recent debates on poverty in New Zealand. The paper then outlines some of the findings in terms of the incidence and severity of poverty in 1993, by household type.

The next steps in the project are to update the data to 1996, evaluate the links between women and poverty, analyse consumer behaviour and link the poverty data with Statistics New Zealand's survey on health status. The objective of the latter is to test whether there is some correlation between an individual's reported age and sex adjusted health status; visitation rates to doctors, hospitals and pharmacies; and the incidence of poverty.

Debates on Poverty in New Zealand

Issues concerning the concept of poverty, its measurement, and trends in the number of people who are poor, have come under media and academic scrutiny over the last few years (Barker, 1996; Easton, 1995; Kerr, 1996; Krishnan, 1995; Stephens et al, 1995; Waldegrave et al, 1996). The Prime-Minister (Mr Bolger) has claimed that nobody in New Zealand is starving, implicitly using an absolute, destitution-based poverty level applicable to less-developed countries.[1] Ms Shipley (Minister of Health) prefers a poverty line of 50% of median income, claiming that this measure is commonly used in international comparisons of poverty.[2] This is close to the current unemployment benefit level, which the Minister describes as a "modest safety net", but is in fact similar to the subsistence goal rejected by the Royal Commission on Social Security (1972).

Krishnan (1995) uses the "benefit datum line" as a standard to evaluate other poverty measures. Easton (1995) compares his benefit datum level, which was based on the work of the Royal Commission of Social Security (1972), with several alternative approaches. Whilst the Commission's determination of a benefit level was guided by substantial research and community submissions, as Easton admits, its "judgement could be wrong – certainly it was not precise". However, the benefit level is no longer based on the 1972 concept, and Easton's (1995) subsequent updating of the 1972 benefit level by the consumer price index means that a relative measure of poverty is being updated by an absolute poverty-based standard. Moreover, the approach ignores the significant social and economic changes that have occurred over the last 20 years, as well as the policy changes which have occurred in the interim period. Updating by mean equivalent disposable income maintains the logic of the Commission's approach, but the starting point is still the 1972 conditions, policies and views, not those of the 1990s (Easton, 1995).

Barker (1996) and Kerr (1996) are more interested in poverty dynamics, claiming that most people are poor for a short period of time. They fail to note that as some people move out of poverty, others move into it. When they use a fixed poverty measure of 60% median equivalent household disposable income through time, they find that there has been a downward trend in the incidence of poverty since 1988. Easton (1996) claims that this result is counter-intuitive, arguing that if a fixed real income poverty line was used, the increase in income inequality and fall in living standards over this period should show an increase in poverty. Barker (1996) also uses an absolute income standard, based on Easton's methodology, using both the 1994 unemployment benefit level and 1994 invalids benefit level, to show that the percentage of households below these standards has fallen since 1992.

The problem with all of these measures of poverty is that they are arbitrary, in the sense that they have no direct relationship with the standard of living that can be achieved.[3] They are thus more measures of income distribution than income adequacy. What is required is a measure which is based on the experiences and expectations of those who live on relatively low income, and which is related to living conditions, policy parameters and social attitudes of the 1990s rather than the early 1970s, or those in other countries.

The major catalyst for the renewed interest in poverty has been the impact of a decade of economic restructuring combined with the reformulation of the welfare state. No longer is there a basically universal welfare state, operated on the principles of 'belonging to and participating in the community', but one targeted on the basis of need and family income, providing only a minimum income. The living standards of beneficiaries and low-wage households have been adversely affected, until 1993, by increases in unemployment and its longer duration, an average cut of 14% in the real level of the social security benefits, higher net housing costs for those in state housing and the negative impact of the Employment Contract Act 1991 for those on low wages. Economic growth and the reduction in unemployment since 1993 should result in a lower poverty incidence.[4]

Establishing a Poverty Measure through the Use of Focus Groups

The methodology of the New Zealand Poverty Measurement project is to combine a household-based micro analysis of income adequacy with a statistical analysis of unit record data from Statistics New Zealand's Household Economic Survey (HES) to give information on the incidence and severity of poverty. In the absence of the micro analysis, which anchors the poverty line within the experience of those who live on low and inadequate incomes, any poverty line is arbitrary. Focus groups provide a consensual poverty threshold, through a series of meetings with low income families, during which they develop an interactive view as to the level of expenditure required to meet a defined standard of living.

The focus groups are made up of 8-12 householders from the same cultural background, family type or labour market category, with a variety of one- and two-parent households, income sources, age and number of children, and housing tenure arrangements. Each focus group develops two financial estimates, one for an expenditure that is "fair for households to participate adequately in their community", and the other for a "minimum adequate household expenditure". It is the latter which is used for the poverty threshold, and requires that the household has a basic set of appliances and furniture etc., no significant costs associated with sickness or disability, and an ability to manage the household's limited budget extremely well.

Table 7.1 shows the results from a sample of the early set of focus groups, undertaken in Lower Hutt, including Wainuiomata, in 1993. The minimum adequate weekly expenditure for two adults and three children was $471, with a relatively small variation of $442 to $491. There are differences in the composition of expenditure between the focus groups, with Samoan and single-parent families having larger food expenditures, and housing costs greater for Samoan and middle income Pakeha families. Clothing estimates were higher for Maori households and transport costs for wage earners. Samoan households spent more on the 'exceptional' category because of donations to the church and family obligations, but much less for recreational activities, while Maori provided for extended family obligations and tangihanga.

This form of budget standards methodology (Saunders, 1996) has the advantage that the independently derived focus group estimates can be verified against other sources of information – housing costs can be checked against Housing New Zealand's data on low cost rents, and food expenditures against nutritional requirements. Differences in the composition of the total budget reflect the cultural attitudes and lifestyles of the various community and householder groups. The consistency between the groups indicated a common view of budget reality of those living on relatively low incomes, and thus helps set a standard of living below which households should not fall. This represents an absolute standard of living, but set relative to the economic and social conditions of New Zealand.

A valid criticism is that the results are Wellington-based. Earlier groups were run in Porirua, and since then further focus groups have been undertaken in rural New Zealand, middle-sized cities, Auckland, and further groups in Wellington (Waldegrave et al, 1996). In the first three columns

Table 7.1: Composition of Expenditure for 'Minimum Adequate Household Expenditure': 2 Adults and 3 Children. Lower Hutt, 1993

Expenditure Category	Maori	Samoan	Pakeha Low Income	Single Parent	Low Wage Earning	Pakeha Mid Income	Average
Food	100	150	100	130	90	100	112
Household operation	10	10	10	15	10	10	11
Housing	150	180	150	150	150	180	160
Power	30	20	20	25	20	15	22
Phone	11	10	10	10	10	10	10
Transport	40	30	40	55	60	50	46
Activities	15	10	25	21	30	20	20
Insurance	11	11	15	20	15	12	14
Life insurance	20	10	20	10	5	25	15
Exceptional	10	20	10	10	5	10	11
Appliances	10	6	4	5	10	5	7
Furniture	10	6	5	0	3	5	5
Medical	15	5	15	5	15	5	10
Clothing	37	10	15	20	20	20	20
Education	6	5	8	15	10	5	8
TOTAL	475	483	458	491	442	472	471

of Table 7.2, a sample of the 1995 rural results is shown. They give a $57 per week lower estimate than for the 1993 Wellington groups. There is relatively little difference between the results for each rural community. The use of expenditure categories means that the source of the differences can be easily identified. The horticultural base for Otaki probably explains the $50 per week lower estimate on food expenditure (the group also described an active barter economy in the area). Otherwise the expenditures are fairly similar to the 1993 Lower Hutt estimates (Table 7.1). The $20 per week lower housing costs, the more stringent approach to exceptional categories and the lower transport costs is all that puts them out of the lower range of the 1993 city groups, despite higher estimates in the power/heating category.

The last three columns of Table 7.2 indicate that the average total expenditure for the Auckland focus groups was $638 in 1996. Two other focus groups were also undertaken in Wellington in the same year and they had an average of $534. Average housing costs in Auckland were $285, and in Wellington $188, in part a product of Auckland's recent property boom. Food costs were $25 per week greater in Auckland, but this was offset by lower activities and recreational expenditure. Other expenditures were fairly similar. Thus the major difference in expenditure between all the focus groups is housing costs, with the other expenditures generally being similar across regions. The results are now a summary of the views of over 400 New Zealand households. The original Wellington poverty line estimates would thus appear to be about average for New Zealand, and a realistic estimate of the minimum adequate budget required by households to avoid poverty and hardship.

Measuring the Incidence and Severity of Poverty

The focus group estimate of the minimum adequate income required for two adults and three children for 1993 was used as the basis for setting the poverty threshold. Stephens et al (1995) and Waldegrave et al (1996) go through the steps of converting that poverty threshold into a poverty line suitable for all family types. As an indication of where the poverty line fitted into the general income distribution, and as the basis for a relative measure of poverty, the threshold was converted into a percent of median equivalent household disposable income.[5] So far the results indicate that 60% of median equivalent household disposable income is appropriate. This is a little below the benchmark for the larger urban areas, and a little

Table 7.2: Composition of Expenditure for 'Minimum Adequate Household Expenditure': 2 Adults and 3 Children. 1995 Rural and 1996 Auckland Focus Groups

Expenditure Category	1995 Rural			1996 Auckland		
	Masterton	Carterton	Otaki	Maori	Samoan	Pakeha
Food	120	120	70	180	150	150
Household operation	12	10	8	10	30	40
Housing	140	120	150	300	275	280
Power	28	46	35	25	30	23
Phone	18	10	10	10	10	10
Transport	25	30	30	40	15	40
Activities	45	20	15	10	-	-
Insurance	10	3	10	-	-	-
Life insurance	-	8	-	-	-	-
Exceptional	-	-	-	20	50	-
Appliances	8	10	15	10	10	13
Furniture	-	-	5	10	15	-
Medical	5	25	10	15	20	10
Clothing	14	15	20	10	30	20
Education	6	4	14	10	15	20
TOTAL	430	421	392	650	645	620

above that indicated by small centres and towns. The benchmark is not a fixed percentage, however, as it will vary with changing economic and social conditions and policy parameters.

Table 7.3 reports on the incidence and severity of poverty in New Zealand for 1993. At the focus group determined poverty level of 60% of median equivalent household disposable income, some 10.8% of households, comprising 13.4% of the population, were poor. After adjusting for housing costs, there was a substantial rise in the incidence of poverty, with 18.5% of households and 20.5% of the population falling below the standard.[6] Many households with relatively low equivalent incomes have above average housing expenditures. While some of this represents a deliberate choice made by young couples taking out mortgages on the basis of lifetime, rather than current income, much of the increase in poverty incidence after housing costs is due to the move to market rentals for state housing prior to the introduction of the accommodation supplement. A sensitivity analysis has also been undertaken by using a lower poverty line, set at 50% of median equivalent household disposable income, in line with the mechanism for setting the post-1991 unemployment benefit level. Only 4.3% of households and 5.5% of the population were below this standard.

The columns 'Poverty Reduction Efficiency' (PRE) indicate the effectiveness of social security benefits in reducing the incidence of poverty, assuming no behavioural responses from the transfer payments.[7] However, when a programme (e.g. pensions) has operated for some time, it will be taken into account when planning future income needs. In the absence of social security payments, 40.3% of households would have been poor, but this fell to 10.8% after receipt of social security benefits, giving a poverty reduction efficiency ratio of 73.2%. The pre-transfer poverty incidence for people is lower at 36%, but the post-transfer poverty rate is higher, giving a lower efficiency estimate. This indicates that large families have better access to market income, but receive less assistance from the state. At the 50% 'safety net', social security payments are naturally more effective, reducing the household poverty incidence from 36.6% to 4.3%. After adjusting the poverty measure for housing costs, the poverty reduction efficiency estimates are lower.

At the focus group determined poverty measure, the mean poverty gap is 15.8% of the poverty line, or $50 per week for a couple with one child. The total poverty gap is $308 million, or just 0.4% of GDP. After

Table 7.3: Incidence and Severity of Poverty, 1993

Poverty Measure	Poverty Incidence		Poverty Reduction Efficiency		Poverty Gap	
	Household	People	Household	People	Mean % Poverty Line	Total Equivalent
50% Income	4.3	5.5	88.2	82.4	13.6	87.26
60% Income	10.8	13.4	73.2	68.4	15.8	308.51
			After Adjusting for Housing Costs			
50% Income	11.5	13.3	71.1	61.7	31.6	454.18
60% Income	18.5	20.5	58.1	51.3	29.7	826.45

Source: Derived from Department of Statistics, 1994.

adjusting for housing costs, both the mean and total poverty gaps are much larger, rising to 30% of the poverty line, or $826 million – 8.2% of social security expenditure and 1.09% of GDP. This poverty gap estimate means that if resources are targeted to those in need, and the source of their need, then benefit levels (including family support) would be raised by $308 million, and housing assistance, through the Accommodation Supplement, a further $518 million.[8] At the lower poverty level, the total poverty gap is $87 million, but the income of the poor is 13.6% below that poverty line. After adjusting for housing costs the total poverty gap rises to $454 million.

Who are the Poor?
At the focus group determined poverty level, Table 7.4 shows that 46% of sole parents are poor, and over 72% after adjusting for housing costs. Most sole parents are not in the full-time labour force, giving an incidence of poverty of 87% before government transfers. The social security system is relatively inefficient at reducing their incidence of poverty, with a 46.6% PRE before adjusting for housing, and a very low 18% after adjustments for housing.[9] The structure of poverty indicates that sole parents account for 22.8% of the total poor, and 20% of the total poverty gap.

Single adults have a relatively low incidence of poverty, especially before housing cost adjustments, but account for about 18% of the total poor (and poverty gap). The poverty incidence for two adults is even lower, and PRE is even higher. They account for just over 10% of the total poor and total poverty gap. There are two groups represented here. Pensioners over 65 have a PRE of 95.6%, giving a post-transfer poverty rate of 3.5%. Those under 60, where the allegiance to the workforce is high, have a low pre-tax poverty rate, but only a 46.6% PRE, resulting in a post-transfer poverty rate of about 7%.

The incidence of poverty is highest for those with children, with the incidence increasing with number of children due in part to the lack of generosity in New Zealand's assistance to low income and large families (Stephens and Bradshaw, 1995). This accounts for the efficiency of the social security system falling as the number of children increase. It is a very low 27% for 2 adults with 3 or more children before adjusting for housing. The incidence of poverty is high also because many of those with large families are Maori and Pacific Islanders where the poverty incidence

Table 7.4: Incidence of Poverty, and Structure of Poverty and Poverty Gaps by Household Type, 1993. Focus Group determined Poverty Level

Household Type	Before Housing Cost Adjustments				After Housing Cost Adjustments			
	Incidence %	PRE*	Structure Poverty	Pov. Gap	Incidence %	PRE*	Structure Poverty	Pov. Gap
1 adult	9.1	83.4	17.4	18.0	19.5	73.7	21.8	20.3
1 adult+children	46.2	46.6	22.8	20.0	72.6	18.3	21.0	20.0
2 adults	3.7	88.6	10.7	11.6	7.6	79.6	12.9	12.9
2 adults+1 child	14.0	54.1	8.1	5.9	24.9	31.0	8.5	7.7
2 adults+2 child	12.4	49.8	11.2	10.8	21.2	29.2	11.2	14.7
2 adults+3+child	24.1	27.2	15.7	18.1	30.3	19.4	11.6	12.1
3+adults	4.2	77.1	4.5	4.7	8.6	60.5	5.4	4.6
3+adults+child	13.3	55.1	9.6	10.9	18.2	45.2	7.6	7.7
Total	10.8	73.2	100.0	100.0	18.5	58.1	100.0	100.0

* Poverty Reduction Efficiency.
Source: Derived from HES (Department of Statistics, 1993).

is greater than average (Stephens et al, 1995). The mean poverty gap also rises with family size, being less than average for couples with one child, but well above average for couples with three or more children. Large families account for a greater proportion of the total poverty gap than the total poor population.

Measuring trends in the incidence of poverty over the period of economic and social reform is heavily dependent upon the measure of poverty, and how it is updated with changing economic conditions. The focus group perception as to what would be a minimum adequate household expenditure cannot be backcast. Real median equivalent household disposable income fell by 17.1% between 1984 and 1993, and mean income by 5.4% (Stephens et al, 1995). The fall has been greatest at the bottom of the income distribution, and least at the top, with only the top 10% of households actually being better off. The Gini co-efficient measure of inequality rose from 0.255 to 0.303, indicating a substantial widening of the income distribution.

If one takes a relative view of poverty, using median income as the base line, then the incidence of poverty, before adjusting for housing, fell from 13.7% in 1984 to 10.8% in 1993. This fall is largely due to the relative rise in the level of the old age pension, going from marginally below the poverty line in 1984 to marginally above it in 1993. The poverty incidence for single adults fell from 27.5% to 9.1% over this period, and two adults go from 6.4% to 3.7%. But all other household groups actually have an increase in their incidence of poverty. After adjusting for housing costs, the overall incidence rises from 14.0% to 18.5%, indicating a relative rise in housing expenditures for low income households. All household types except single adults have an increase, but the largest increase is for sole parents, from 48.8% to 72.6%. It is thus very dangerous just to look at the overall incidence of poverty and draw conclusions as Barker (1996) and Kerr (1996) do – it is necessary to look at the detailed data to ascertain why those changes have occurred.

If an absolute poverty standard is taken, then a substantial increase in the incidence of poverty over the 10-year period is shown, more than doubling from 4.3% in 1984 to 10.8% in 1993. The increase occurs for all household types, with the most dramatic change being for sole parents from 11.8% to 46.2%. The incidence for single adults went from 3.7% to 9.1%, and for families of two adults and three or more children from 14% to 24.1%. After housing costs, on the absolute poverty measure, there has been a trebling of the poverty incidence.

Both the absolute and relative incidence figures must be treated with caution as they do not derive from the focus group methodology, and thus do not take account of how the poverty line would change for the economic and social upheavals of the last decade. The impact of policy changes such as user charges or GST, and the population's perceptions as to what constitutes a minimum adequate income, cannot be cast backwards by the focus groups. Since 1993, they have been continuously updated. The percentage of median income will alter with lower unemployment, tax rate changes and different economic conditions.

Conclusions

This paper has given an indication as to the analysis so far undertaken by the New Zealand Poverty Measurement Project team. Future work includes linking the macro poverty analysis with the HES survey on health status to see if there are any correlations between these variables, to continue updating the focus group work and macro measures of poverty incidence and severity, and investigating the relationship between women and poverty.

The focus group offers a robust method for setting a consensual poverty line. It shows that poverty measures are time specific, depending upon current social attitudes, economic conditions and policy parameters. The macro data provide statistical information on the incidence and structure of poverty, by a variety of ways of categorising household types, the severity or depth of poverty, and the effectiveness of social security benefits in reducing poverty. Adjusting for housing costs permits an analysis of the effectiveness of the shift to market rents and the targeted accommodation supplement. The research has shown that poverty is a significant problem in New Zealand, especially for particular household groups and ethnic categories, and that the social security system is effective in reducing the incidence of poverty among pensioners, but not for those with large dependent families.

Acknowledgements

The research has been funded by several grants from the Foundation for Research, Science and Technology, organised through BERL by Paul Frater and the Family Centre by Charles Waldegrave. Statistical analysis from the Household Economic Survey has been undertaken by Statistics New Zealand.

References

Barker, G (1996) 'Poverty on the decline', *City Voice*, 17 October

Easton, B (1976) 'Poverty in New Zealand: estimates and reflections', *Political Science*, 28 (2)

Easton, B (1995) 'Poverty in New Zealand: 1981-1993', *New Zealand Sociology*, 10 (2), November

Easton, B (1996) 'Poverty and inequality', *City Voice*, 10 October

Eurostat (1990) *Poverty in Europe: the Facts,* Brussels: European Commission

Jensen, J (1988) 'Income equivalences and the estimation of family expenditures on children', mimeo, Wellington: Department of Social Welfare

Kerr, R (1996) 'Sharing economic gains', *Evening Post,* 23 September

Krishnan, V (1995) 'Modest but adequate: an appraisal of changing household income circumstances in New Zealand', *Social Policy Journal of New Zealand,* 4, July

Mitchell, D (1990) *Income Transfers in Ten Welfare States,* Aldershot: Academic Publishing

Reddy, C (1996) 'Measuring poverty differently', *The Hindu,* 22 July

Royal Commission on Social Security (1972) *Social Security in New Zealand,* Wellington: Government Printer

Saunders, P (1996) 'Development of indicative budget standards for Australia: project outline and research methods', BSU Working Paper Series, No. 1, September, Budget Standards Unit, University of New South Wales

Stephens, R (1994) 'The incidence and severity of poverty in New Zealand, 1990-91', GSBGM Working Paper 12/94, Victoria University of Wellington

Stephens, R and J Bradshaw (1995) 'The generosity of New Zealand's assistance to families with dependent children', *Social Policy Journal of New Zealand,* 4, July

Stephens, R, C Waldegrave and P Frater (1995) 'Measuring poverty in New Zealand', *Social Policy Journal of New Zealand,* 5, December

Waldegrave, C, S Stuart and R Stephens (1996) 'Participation in poverty research: drawing on the knowledge of low income householders to establish an appropriate measure for monitoring social policy impacts', *Social Policy Journal of New Zealand,* 7, December

Notes

1   Reddy (1996) commented that "for most people [in India] poverty is a relative concept. At the extreme, those who go hungry every day, do not have adequate clothing and sleep on the pavement are clearly poor. But families which do not suffer from malnutrition and yet cannot afford to buy medicines when illness strikes will also, rightfully, consider themselves poor".

2   The Luxembourg Income Study uses 40%, 50% and 60% of median equivalent household disposable income as their poverty standards, with more generous equivalence scales than are commonly used in New Zealand (Mitchell, 1990). The European Community uses 40% and 50% of mean equivalent household disposable expenditure, with similar equivalence scales to Jensen (1988) (Eurostat, 1990). In New Zealand in 1991, 50% mean expenditure was almost identical to 60% median income (Stephens, 1994).

3   When originally set, Easton's (1976) benefit datum line, based on the Royal Commission of Social Security's (1972) recommendations, was an exception. Its relationship to current living standards is far more problematic.

4   The recent Auckland property boom, forcing up housing costs for low income households, may have adversely affected the incidence of poverty after housing costs.

5   Using the Jensen (1988) equivalence scales for both determining median equivalent household disposable income and for setting poverty levels for other household types. Outliers, defined as those declaring self-employed losses or with expenditure three times their income, were excluded.

6   The approach taken here is to exclude all housing-related expenditure from each household's income, and then recalculate median income and set the new poverty threshold at 60% of median income less housing costs. An alternative approach, suggested by George Barker, was to use the focus group minimum adequate expenditure less housing costs as the after-housing cost poverty line. This gives a lower poverty incidence of 11.9% of households (and a lower poverty gap of $369 million). This approach originally was rejected as it would keep, *ceteris paribus*, the poverty incidence of state housing tenants constant when state housing rents were rising to market rents, but the Barker hypothesis is being tested.

7   The formula is based on after-tax income, and is calculated as (Incidence before social security benefits less Incidence after receipt of social security benefits) divided by Incidence before social security benefits.

8   Perfect targeting would not be expected due to difficulties of precisely ascertaining individual circumstances, and would not be desirable due to adverse labour supply incentive effects of the implied 100% effective marginal tax rate.

9   The separate impact of the accommodation benefit has not been calculated, but very few people in the HES data base report receiving the accommodation benefit.

8 ~ Socioeconomic Inequalities and Health

Commentary 1: Cynthia Kiro

I have been asked to comment on papers presented by both Robert Stephens and Charles Waldegrave, and George Barker.

Stephens and Waldegrave's paper focuses on the measurement of poverty in New Zealand and seeks to demonstrate that significant poverty does exist whatever standard measure one chooses to use, such as 50 or 60% of median income, relative or absolute poverty.

There have been many criticisms since their earlier effort, primarily around the methodological limitations of their approach which is focus group-oriented and where the poverty line is self-defined. While there remain methodological flaws in their current approach, this still represents a significant contribution to a reasonable understanding about a poverty threshold, and the nature and extent of poverty in our society.

In the absence of an official government poverty line, this represents a much needed contribution. The difficulty is that this paper does not address one of the main aims of the Conference, namely to apply these findings to policy on health. They do signal their intention to apply their findings to health – but this is in the future.

The most useful element of this paper is that we gain considerable insight into the priorities of different households according to geographical location (with the expansion of their study to include rural and another metropolitan centre), culture (Maori, Pacific Island and Pakeha families), sole parent and two parent households and single adult households.

Their assignment is that we need to understand the nature of these households and how they assign value to various items like housing, food, clothing, church and family commitments etc. to comprehend the effects of various social policies and our response to them.

Despite many criticisms, primarily methodological, which were aimed at their previous work with Frater, Stephens and Waldegrave provide a significant grounding to the debate about the worsening experience of poverty for some New Zealanders. This is consistent with international studies showing an increasing gap between rich and poor.

121

Earlier this year during the media campaign around the Children's Coalition Conference on the Multiple Effects of Poverty on Children and Young People, it was notable that the Prime Minister, Jim Bolger, and Minister of Health, Jenny Shipley, responded to the poverty debate by denying that poverty existed in New Zealand. This position shifted quickly to a discussion about the definition of poverty – whether the poverty line should be set at 50-60% of the median income, and whether there was relative or absolute poverty.

This paper is a defence of these earlier criticisms from academics, researchers and politicians with a considerable expansion in the number of focus groups. The main message of this paper is that no matter what basis you use to measure poverty (comparable to international standards), it does exist in good measure in New Zealand, and it is worsening for some households.

The paper also raises questions about the supposed transitory nature of poverty advocated by Barker. Rather, they claim that as some poor people move out of poverty (below the poverty threshold), then others move in to take their place. In discussion, Waldegrave made the point that we may expect to see those closest to this threshold moving in and out of the threshold, depending on what is happening in their lives, e.g. child-bearing, relationship break-ups, purchase of own home etc.

Stephens and Waldegrave include tables which adjust for social security benefits and housing, the latter representing the single largest household item of expenditure. It is clear from this analysis that in those areas where housing costs are high, there is a major impact on household expenditure. More could have been made of the figures for Auckland, however, given the difference of around $100 for these households compared to other parts of the country.

Stephens and Waldegrave conclude from these findings that "the move to market rentals from state housing prior to the introduction of the accommodation supplement" is responsible for much of the increase in poverty incidence. The conclusions about who are most affected by the worsening incidence of poverty are compelling. These groups are those with children, especially large families, sole parents and Maori and Pacific peoples.

We must acknowledge that these groups are the deliberate targets of ongoing social policy decisions. The outcomes for these groups of people are part of a value-based process by which policy decisions are made and implemented, since social policy reflects streams of value decisions and implicit priorities.

An old morality is at work here – and it should be unmasked. There is little room in our brave new world for large families, sole parents (overwhelmingly consisting of women and children), and the communally-based value systems and cultural practices of Maori and Pacific peoples.

These results have important implications for policy-makers in health in prioritising the health needs of most at-risk populations. Notably, changes in these outcomes will require policy interventions which are just as deliberate with a view to the long-term rather than immediate political cycles.

There is an evident incongruence between the macro-economic and social policies pursued and the health problems being experienced by vulnerable populations within Regional Health Authorities areas. To a large extent health agencies must pick up the pieces of these macro interventions which assign certain groups of our society to the impoverished underclass. This is especially noticeable for children – who are most innocent yet vulnerable to such changes.

Intuitively we know that having insufficient money to live a reasonable lifestyle within society is bad for health, significantly in the accumulation of stress, compromised housing conditions, poor nutrition, limited life experiences and so forth. This is no gateway of opportunity for upward mobility, despite considerable effort from those below the poverty line to hold their lives, and those of their children, together. Oakley wrote in 1992 "... stressful life events and circumstance are unequally distributed among social class groups. Generally, those of lower classes experience more stressful life events and circumstances".

George Barker's paper is of considerable length and presents a defence of some of the criticisms of New Right policies and a model to explain why people should support such a rational approach to social policy. His main point seems to be that much rests on the idea that given the right inducements, people will move out of poverty. Many would agree with this view. Questions arise in considering what those inducements and the other environmental factors might be.

Barker attempts to demonstrate that social policy (in health) should be subject to the same rationality as in other areas, and that coordinated interventions of the private, voluntary and state sectors are needed. Moreover, he identifies the role of the state as being that of a final arbiter between contractual parties with competing interests.

However, despite this rehashing of neo-liberalism and New Right thinking, the evidence provided is very light. For example, his argument about mobility – people moving out of poverty – is based on US data. This is a highly theoretical piece of work which requires much more grounding in what has happened over the last ten years in New Zealand.

Barker's case for a comparative systems approach is that every area of policy which invites assessing alternative institutional structures according to the process and outcomes they involve seems reasonable until you realise that much of what occurs in social policy arenas cannot be understood within the framework proposed by him. Barker has distilled his view of what ought to be, thus offering a normative view, while espousing greater objectivity on social problems.

It is not clear how this paper relates specifically to health policy, except as a justification for key parts of the post-1991 health reforms, with their emphasis on contracting of private, voluntary and state agencies. The paper fails to provide adequate guidance as to the relationship between various levels of policy – national, regional and local, and how various agents of the state can best address health concerns within these different levels.

To some extent my difficulty in reconciling the disparate views taken by Barker, and Stephens and Waldegrave, reflects the fact that the underlying values and assumptions differ so much. As a result they seek to answer different questions. However, we should be in no doubt of the link between inequalities in health and socioeconomic status. In 1990 Malcolm said:

> Inequalities in health occur as a result of certain social or ethnic groups being exposed to a less healthy environment with regard to work, housing, occupation and education and fewer resources available to adopt healthier life styles or access to health.

Seitz et al wrote (1985):

> ... chronic stress is a significant impediment to effective family functioning and ... poverty both increases the likelihood of such stress and restricts the resources available to families to cope with it.

Much of this work identifying the dynamic of socioeconomic inequality has been contributed by social scientists and is considered well researched within this literature.

The contribution of this Conference is in providing a greater credibility by adopting a more rigorous scientific approach. We must be careful, however, not to allow the work done in measuring and defining poverty to persuade us that we do not need to act. It will not be possible to have ideal information as the basis on which to make relevant policy decisions to improve health outcomes for vulnerable populations. Evidence of the growing gap between rich and poor within New Zealand and overseas and its effects on health requires immediate action to halt these conditions.

References

Malcolm, L et al (1990) *Economics and Organisations: Factors in Achieving a Better Health Future for Maori Women,* Department of Community Health, Wellington School of Medicine

Oakley, Ann (1992) *Social Support and Motherhood: The Natural History of a Research Project,* Oxford: Blackwells

Oakley, Ann (1993) *Essays on Women, Medicine and Health,* Edinburgh: Edinbugh University Press

Seitz, A et al (1985) 'Effects of family support intervention: a ten year follow up', *Journal of Child Development,* 56, pp 376-91

Note

1 George Barker proposes this in his dynamic model of social policy. Stephens and Waldegrave point out that Brian Easton argues that this is counter-intuitive.

Commentary 2: Susan St John

As a policy analyst rather than a poverty specialist, I am most interested in how research informs policy development. I found the Barker paper quite difficult to review in the context of the objectives of this conference. The Stephens and Waldegrave paper is more straightforward. They have developed a methodology for an independent measure of poverty which, at some time in the future, will be related to health status.

Unfortunately if and when this is achieved, we may still not know what to do in a policy sense. If there is a correlation between poverty and ill health, we need a hypothesis to explain the correlation before we can talk about policy implications. The hypothesis that the poor are poor because they are sick would suggest an emphasis on treatments including prevention. Health purchasing for the poor becomes the focus. But what if the poor are sick because they are poor? This hypothesis suggests that health purchasing is at best palliative and that a reduction in the poverty itself is more appropriate. If people are poor and sick because of some third factor such as age, race or individual fault, other policy implications follow.

The hypothesis one accepts is likely more than anything to reflect one's ideological preferences. Economic liberals are much more likely than social liberals to discount the role of government policy both in making the poor, poor, and in its potential to alleviate the problem. I am not optimistic that one side will ever convince the other.

One view of the ideological divide is expressed in Saunders' paper:

> The failure of the social policy research community to be sufficiently professional in much of its work has assisted the economic liberals to take over the field with their own studies, often based on the application of simplistic neoclassical economic models of rational choice which are devoid of institutional understanding and policy detail.

There should be no prizes for which of *these* two papers presents the economic liberal approach! While there may be some grain of truth in the Saunders criticism of the social policy research community at large, I do not believe it applies to research such as that of Stephens and Waldegrave. I think that it is unrealistic to expect that attempts to further refine their

126

measures of income inequality and income deprivation will counter the ascendancy of economic liberalism. More rigour may be applauded by the academics but will it provide more clarity in these muddy waters?

A major problem is the void between what should be measured in theory and what can be done in practice. Thus, while Barker extols a total income concept which includes the imputed or psychic income from owning assets, he proceeds to use examples which confine the concept of income to certain types of monetary income only. Stephens and Waldegrave likewise do not impute income from assets and are forced into making ad hoc adjustments to their income measure afterwards. Principally, they do this via an after-housing costs measure of the poverty line.

Easton (1997) has criticised this approach because it is far from clear that application of equivalence scales which are designed to reflect economies of scale in households of different sizes are valid once housing costs have been excluded. I am raising another objection which is illustrated from the following example. Take a family on $15,000 who live in an old caravan and whose housing is totally unsatisfactory but cheap. After housing-costs they may look much better-off than another family on the same income but who pay $200 a week rent for an adequate home. Both families actually have the same potential consumption or 'total income' and are both equally worse-off when compared to a debt-free owner-occupier family on $15,000.

Now suppose that one of the first households has high health care costs because family members are in poor health. Should we extend the Stephens/ Waldegrave approach and look at income after health care costs? If so, what of the family that underspends on appropriate care?

Alternatively, using the total income approach, should good health be regarded as an asset and an income from being in good health imputed to the second household? Should the idea be extended to other intangible but vital assets? For many of us after all, the tangible assets of house, car and superannuation hardly rate against the assets of one's good health, family and human capital.

Thus an 'independent' measure of poverty which can then be linked to illhealth may be problematic. My point is that one should be careful before invoking a total income concept because to do it properly takes us a very long way from the simplistic income measures that we find in both of these papers.

Barker takes a dynamic rather than a static view of income distribution. By establishing that upward mobility occurs, he implies that the existence of poverty therefore does not matter.

I take from Saunders' paper that inequality as the end product of economic liberalism in turn may provide part of its contradiction. If income inequality produces more illhealth, especially among the poor who have not got access to private healthcare, there will be rising pressures on the state's budget. To that, one might add the fiscal pressures of increased spending because there is increased violence, crime and abuse. The danger is that this may lead to an underclass of poor, and the better-off begin to see the problems of the poor as insoluble and therefore not worth attempting to solve.

It is this kind of dynamic to which I wish Barker would devote some attention. His discovery of the mobility of the lowest income decile of wage and salary earners is hardly surprising as it is well accepted that there are lifecycle influences which make the income distribution of individuals over their lifetime more equal than a snap shot picture. Older people can expect to be downwardly mobile in retirement and younger people moving through education can expect to be upwardly mobile. I take it from the flavour of the paper that this mobility justifies a non-interventionist attitude from the state. It is also in tune with the idea that the poor are poor because they made the wrong choices. Poverty becomes the spur to make sure that they escape the condition and another justification for the government doing nothing about it.

Barker ignores the evident social distress made visible in the explosion in the use of foodbanks, in hospital admissions for childhood diseases, increased youth suicides and poor and overcrowded housing conditions. It is not convincing to leave the impression that the poor are all that way because they have made wrong choices and that mobility will solve their problems.

Economic liberals are in essence not interested in government solutions. Economic growth and economic activity is a rising tide that lifts all boats.

Paul Krugman, a highly influential and respected economist, claims that the most remarkable thing about the debate over inequality in the USA is the failure of conservatives to admit that it happened at all. He concluded his book *Peddling Prosperity* as follows:

> America has two great economic problems: slow growth in
> productivity and rising poverty (which is the consequence

of both inadequate productivity growth and increasing income inequality). Everything else is either of secondary importance or a non issue.

He goes on to explain that governments basically do not know how to solve these problems, and those who think they have solutions have a certainty based on ignorance. Nevertheless, Krugman claims that the government can do many things to diminish these problems:

> Let us spend more on programmes that help poor children, from nutrition and medical programmes for poor mothers up through to aid for distressed school districts. Let us also raise the income support given to poor families (international comparisons show that the main reason the US has so many more poor children than other industrialised countries is that it spends less public money keeping children out of poverty). All of this will cost more money – but not that much more because our poor are so poor that it only takes a moderate amount of spending to make them much better off.

Perhaps, too, in New Zealand it is not too much to ask that the worst features of the poverty explosion be ameliorated, not by more studies of the income distribution but by humane and adequate government expenditure.

References

Easton, B (1997) 'About time we had some facts' (Review of *Children of the Poor* by Mike Moore), *New Zealand Books*, March 1997, pp 14-16

Krugman, P (1995) *Peddling Prosperity*, London: Norton and Company

9 ~ The New Zealand Socioeconomic Index: A Census-based Occupational Scale of Socioeconomic Status

Peter Davis, Philippa Howden-Chapman and Keith McLeod

Introduction

The last decade has seen an unprecedented amount of activity devoted to the restructuring of health services in New Zealand. Issues of public health and health promotion have, by contrast, received much less attention. Yet while those concerned with improving public health have been preoccupied with restructuring, research has been accumulating showing that unemployment, poverty and poor housing have a more significant impact on the public's health than nurses, doctors and hospitals. Despite this, issues of socioeconomic difference have not been well canvassed. For example, recent official reviews of health in New Zealand, *Our Health Our Future* and *A Picture of Health,* while addressing aspects of socioeconomic difference, do not include any comprehensive measure such as one based on occupation.

There are many ways to measure socioeconomic status, but little agreement about what exactly is being measured and its significance (Duke and Edgell, 1987). More reassuringly, there is a high correlation between different categorical measures of SES and health outcomes (Bunker et al, 1989). This consistency suggests that measures of socioeconomic position are tapping deep, social structures that are fundamental to factors promoting health. The concept of SES is predicated on the perception of distinct communities with identifiable sub-cultures and varying economic and political power.

On theoretical and pragmatic grounds, the most common measures of SES are occupational groupings (Najman and Bampton, 1991). As a measure, occupation has the advantage that it can be related to other indicators such as income, education and social status; there is considerable agreement about its measurement; and it clearly links to the economic, social and political structures of society. Pragmatically, occupation is a particularly convenient measure of SES because it is frequently the only relevant information available on official health data and death certificates. Its deficiency lies in the problematic status of those groups which are not in the labour market – housewives, the unemployed, beneficiaries and the retired.

For the last two decades in New Zealand, most of those carrying out research into the impact of socioeconomic conditions in the social sciences have used the Elley-Irving Scale index of male occupations (Elley and Irving, 1972, 1976, 1985) or the Irving-Elley index of female occupations (Irving and Elley, 1977). However, as these scales were last revised using data from the 1981 census, a major update is timely.

Project Aims

This paper reports the preliminary results of a study, funded by the Health Research Council and carried out in co-operation with Statistics New Zealand, to update existing occupational scales of socioeconomic status in New Zealand with a new instrument (the New Zealand Socioeconomic Index [NZSEI]). We used the 1990 revision of the New Zealand Standard Classification of Occupations (NZSCO), where greater attention was paid to the skill requirements of job titles than in the previous 1968 version. For example, manual occupations have been grouped by broad skill requirements into three major groups – Trades, Machine Operators and Elementary Occupations.

A secondary aim of our project was to maintain continuity with existing instruments such as the Elley-Irving Scale and data series. We also wanted to establish and maintain international comparability by applying an acceptable statistical model to the development of the Index.

The results presented at this Conference explore population norms for men and women in the paid workforce, but in the future we plan to extend the analysis to households.

Model

The model used in the development of this scale is developed from the theory of returns to cultural capital. Occupation is seen as an intermediate variable indicating the transformation of cultural capital and resources (education) into material rewards (income).

The model is based on the assumption that the link between formal education and income is mediated by an (as yet) unmeasured variable, occupational SES (see Figure 9.1). Under this model, the scaling of occupation is calibrated simultaneously to maximise both the indirect influence of education on income (β_{32}) and the linkage between occupation and income (β_{43}), while at the same time minimising the direct effect of education on income (β_{42}). A statistical formulation of this relationship enables us to estimate occupational scores representing a hierarchy of education attainment and income. The methodology for this study is based on an iterative algorithm developed by Ganzeboom et al (1992) and uses an alternating least squares method. Ganzeboom and colleagues used this methodology to develop an International Socioeconomic Index of Occupational Status that was derived from data from 70,000 people in 16 diverse countries.

While this relatively simple model provides the starting point for the construction of the scale, there are a number of complicating methodological and conceptual issues. For example, age weakens the relationship between education, occupation and income because, although the level of formal education required for jobs has steadily increased over time, older people tend to have higher incomes, even though they have less formal education. Therefore adjusting for age has to be explicitly introduced into the analysis (β_{21}).

There is another set of issues concerning whether categorical or continuous measures should be used. The advantages of continuous measures are their evident parsimony – a great deal of information is summarised in a single parameter – and their amenability to multivariate statistical analysis. On the other hand, categorical measures have both presentational advantages, because data analysed by key social groups are easily understood by the public, and there are conceptual advantages because the categories evidently correspond with identifiable social groupings in society. In this project we will be developing both sets of information; continuous measures are available for more quantitative techniques and, at the same time, these data will provide the basis for the formation of distinct groupings organised around the major skill categories in the NZSCO system.

Method
Initially, a working set was established on a 10% sample from the 1991 census data and scores derived for the full-time, male, European working

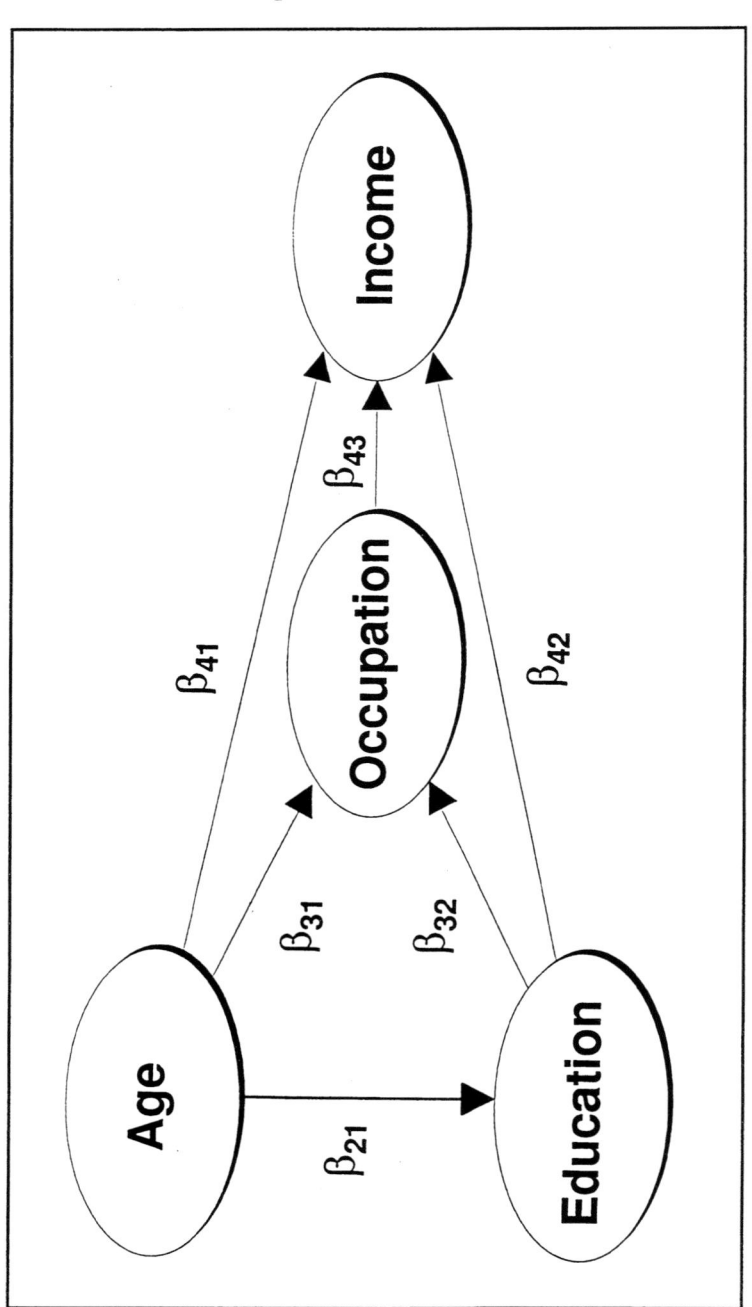

Figure 9.1: Model of NZSEI

population. This was checked against international figures and then the analysis progressively extended to incorporate part-timers, women and non-Europeans. Finally, the analysis was run on the full Census data set. The results presented here are derived from a sub-set drawn from the full Census of over one million people consisting of those in full-time work and aged between 21 and 64.

Occupations with less than three respondents were deleted. Using the NZSCO at the three-digit level means there are 97 occupations. In contrast to the Elley-Irving scale, we incorporated a number of modifications; in particular, age was incorporated into the analysis and special attention has been paid to those occupational groupings showing high levels of employment of women and Maori. In the future we plan to extend our scale to those outside the paid workforce.

Finally, the predictive validity of our Index was tested by assessing its predictive power against various indicators of health behaviour derived from the 1992-3 Household Health Survey.

Results

For convenience, the NZSEI results are given at the one-digit level, so that the occupational groups are ranked categorically from 10 to 90. Sometimes they are presented as unlabelled quintiles rather than class groupings, although at a later stage we plan to reconcile these quintiles with more conventional SES groupings, which match commonly accepted categories. In this section of the paper three sets of results will be presented – basic descriptive information, ethnic and gender differentials, and health indicators.

Descriptive Results

In Figure 9.2, occupational groups at the one-digit (or major group) level are ranked according to their average NZSEI scores. As can be seen, the average SES scores for each occupational grouping are fairly even spread, although the differential between Associate Professionals and Managers is only marginal.

The occupational spread is shown in another way in Figure 9.3 to demonstrate the predictive value of the scale. It shows a 'bubble plot' of smoking levels at the three-digit level of the NZSCO. The logistic regression line for smoking on our Index is superimposed over the data. It shows the expected negative relationship between occupations and smoking.

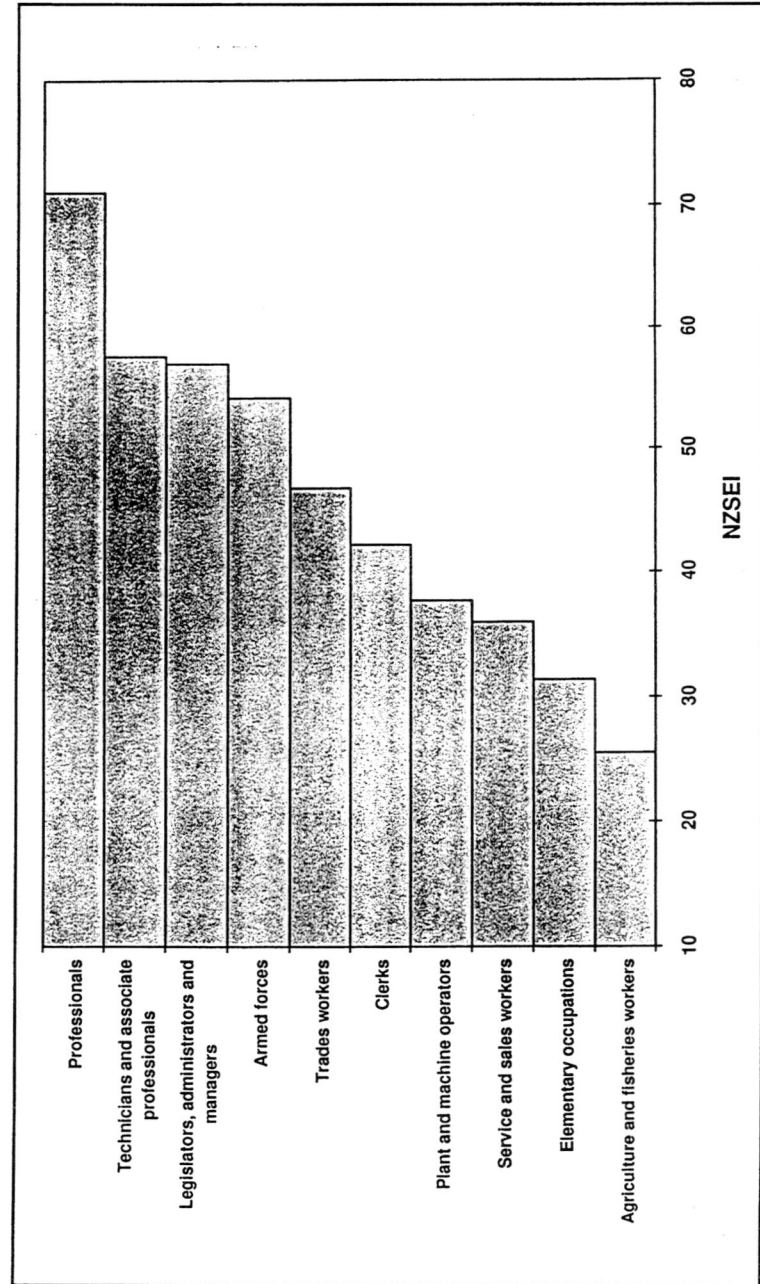

Figure 9.2: Ranking of Occupations by NZSEI

Figure 9.3: Logistic Regression of Smoking on NZSEI

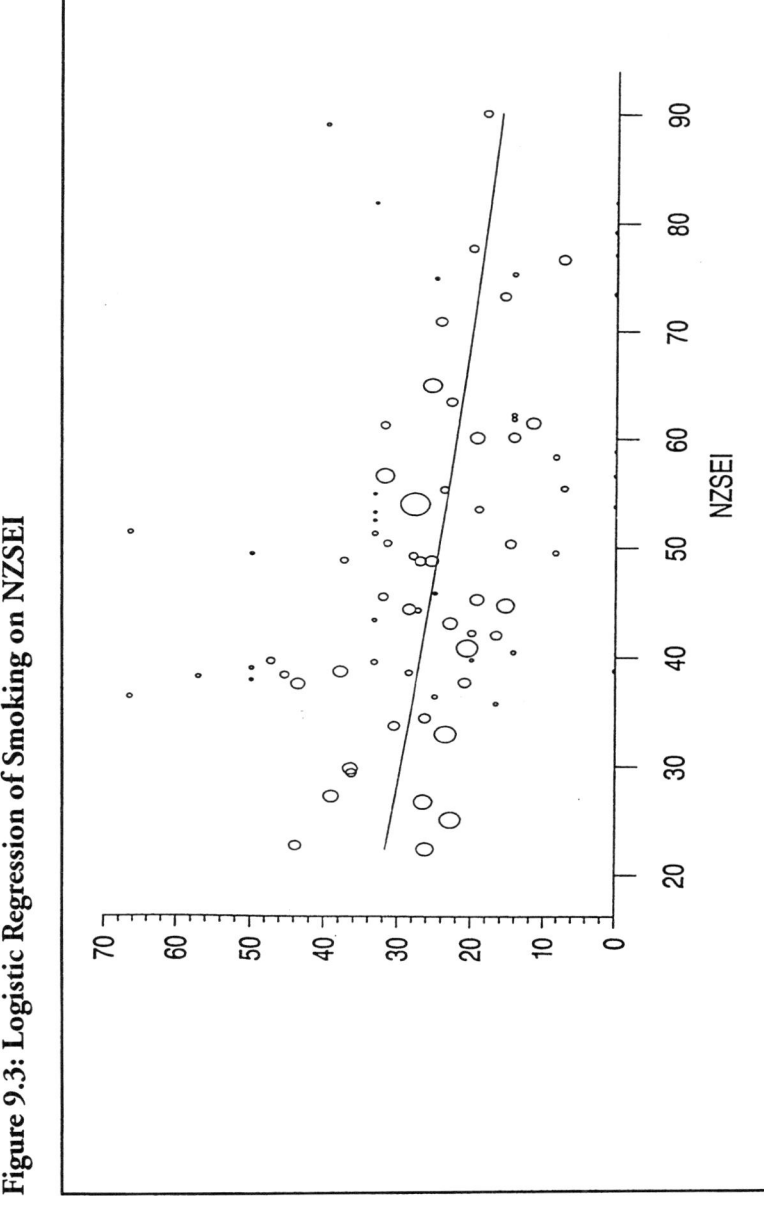

NZSEI

Note: Occupations with only 1 or 2 respondents have been deleted.

Differentials

In Figure 9.4 the average scores for each occupational grouping have been broken down by gender. As can be seen, males have consistently higher SES scores than females in every occupation. This reflects the fact that, in comparison with females in the workforce, males generally have higher incomes, higher levels of educational attainment or a combination of both. The smallest differential is evident in the Armed Forces.

The margin between Maori and European SES occupational scores is also marked (see Figure 9.5). The largest differentials occur in the higher socioeconomic status groups; the gap is less marked in some manual occupations and among Agricultural and Fisheries workers.

Health Indicators

The predictive value of the NZSEI has been tested in a preliminary analysis against a number of health indicators. For example, Figure 9.6 shows that when the percentage of people who assessed their own health as "poor" in the Household Sample Survey is graphed by NZSEI quintile, the expected negative relationship was revealed (that is, higher SES occupations are less likely to rate themselves in poor health). However, this result was not entirely consistent with expectation since there was a departure from a strict gradient in the bottom two quintiles.

Further attenuation of the relationship is revealed once the analysis is adjusted for age, sex and ethnicity; the bottom three quintiles now show little differentiation (see Figure 9.7).

The pattern apparent in Figure 9.6 is replicated in Figure 9.8 which shows the percentage of people smoking by Index quintiles. People in the lowest quintile smoked more (25%) than those in the highest status quintile (16%), but the highest proportion of smokers (32%) was in the second-to-bottom quintile. These relationships were unchanged when the proportion of smokers was adjusted for age, sex and ethnicity.

Despite these apparent departures from strict linearity in the socioeconomic gradient, overall the expected association between lower SES and poorer self-assessed health and greater levels of smoking is confirmed. The Index was also regressed against a chronic illness, asthma, where there is no evidence that SES is associated with prevalence (though there is evidence that it is related to severity). These results are presented in Figure 9.9, which indicates no particular pattern, with the third quintile having a slightly higher proportion (11%) of people with asthma than the other

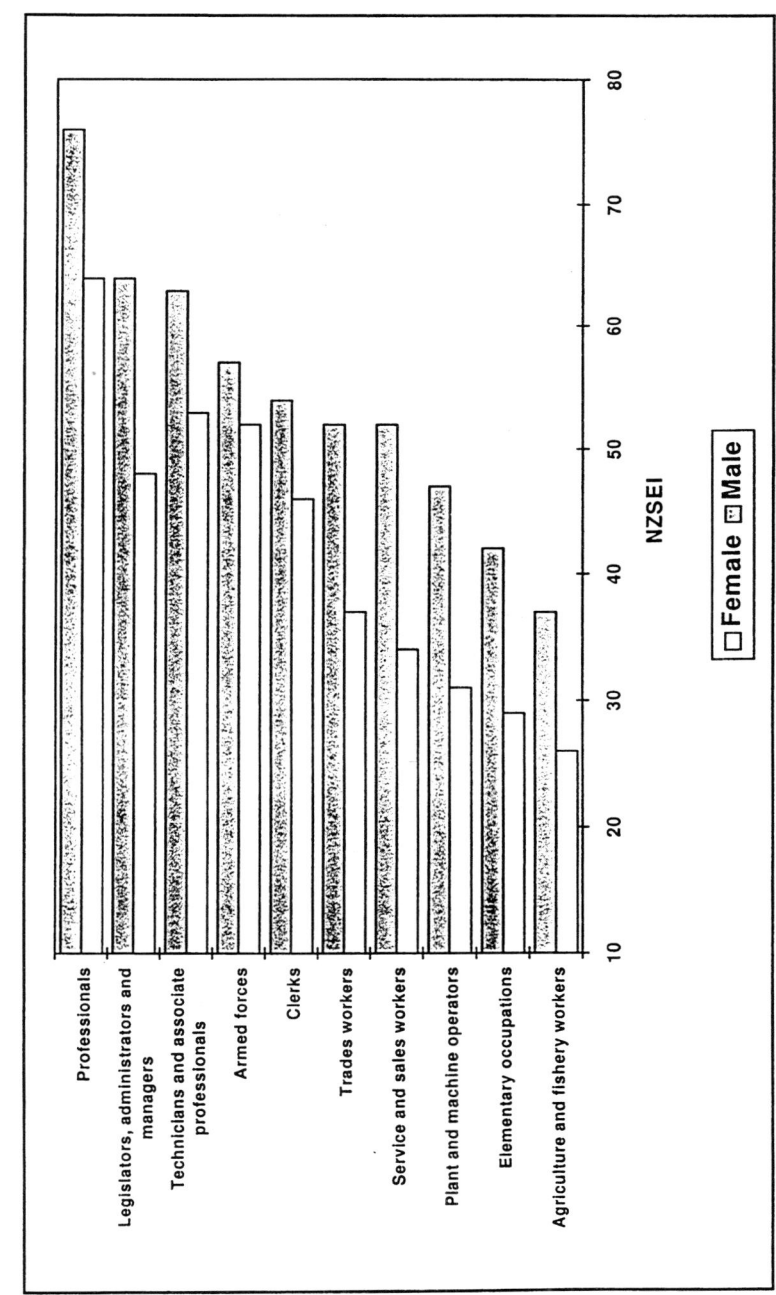

Figure 9.4: Male-Female Differentials in NZSEI Scores

Figure 9.5: Maori-European Differentials in NZSEI Scores

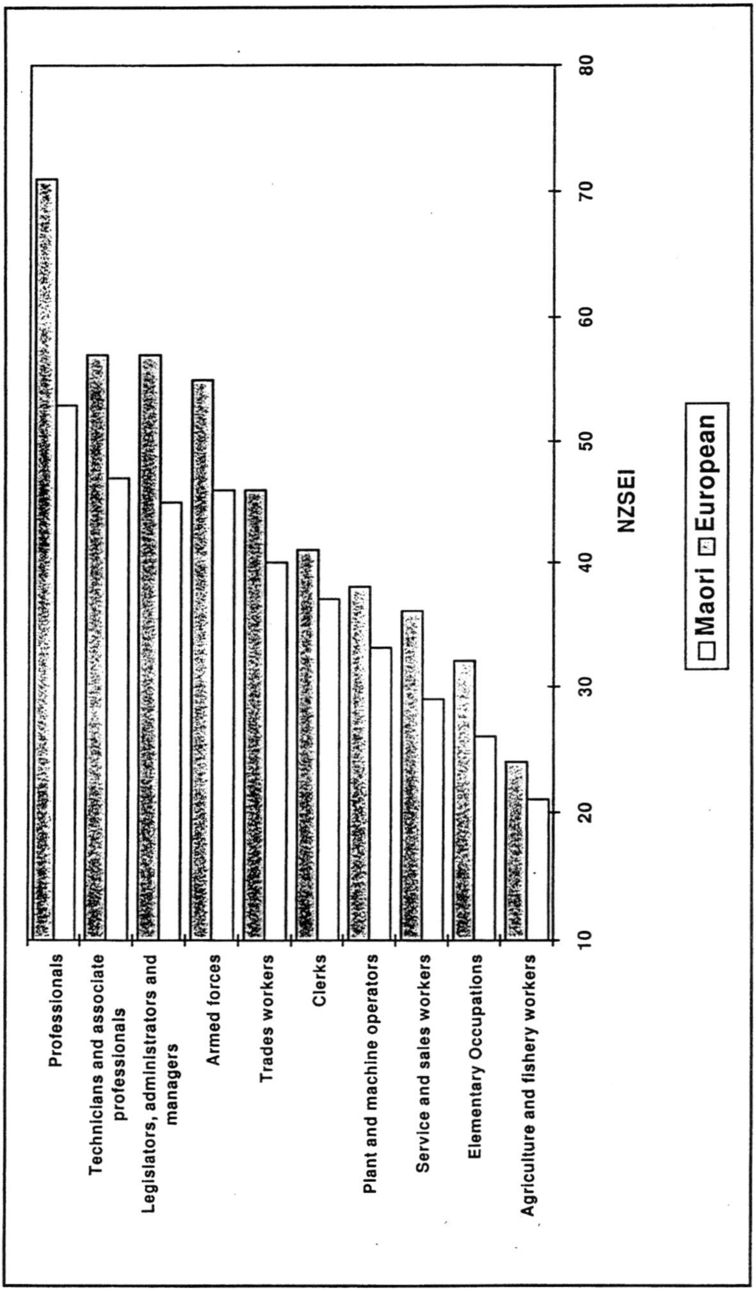

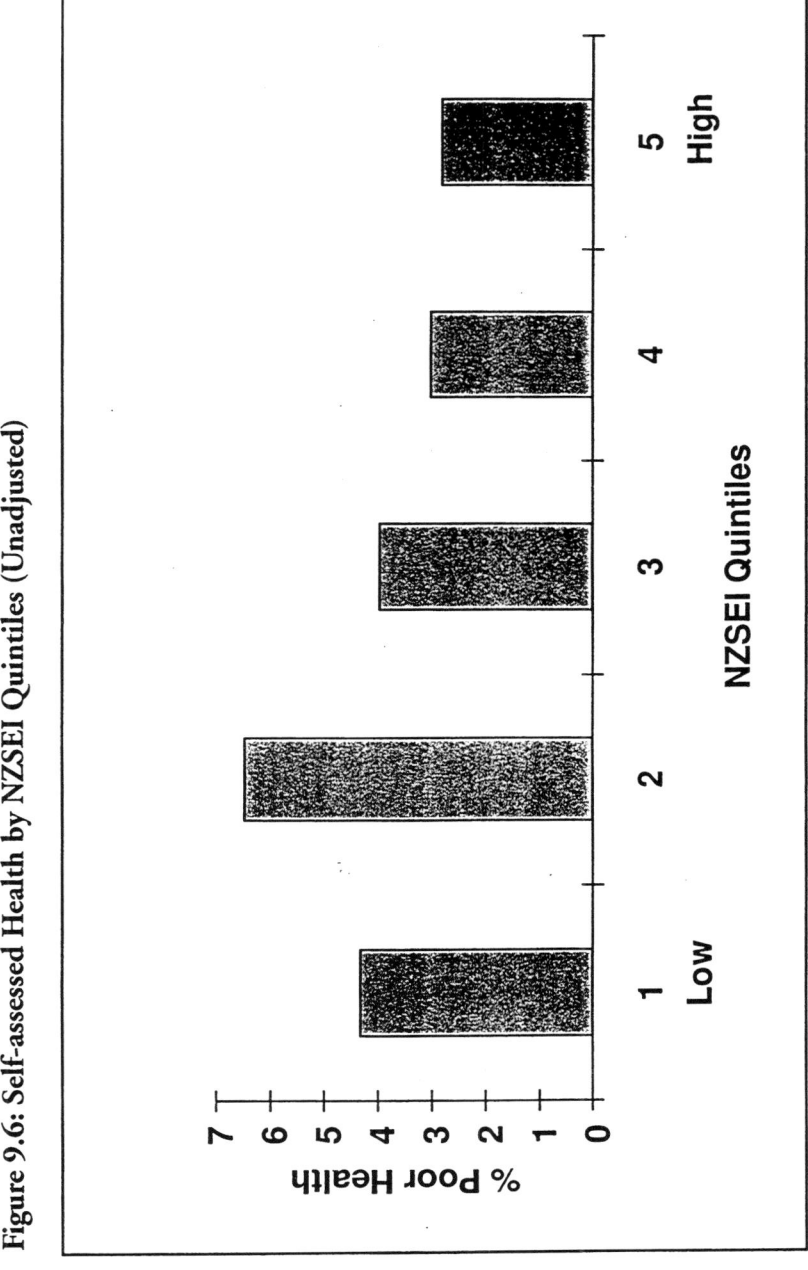

Figure 9.6: Self-assessed Health by NZSEI Quintiles (Unadjusted)

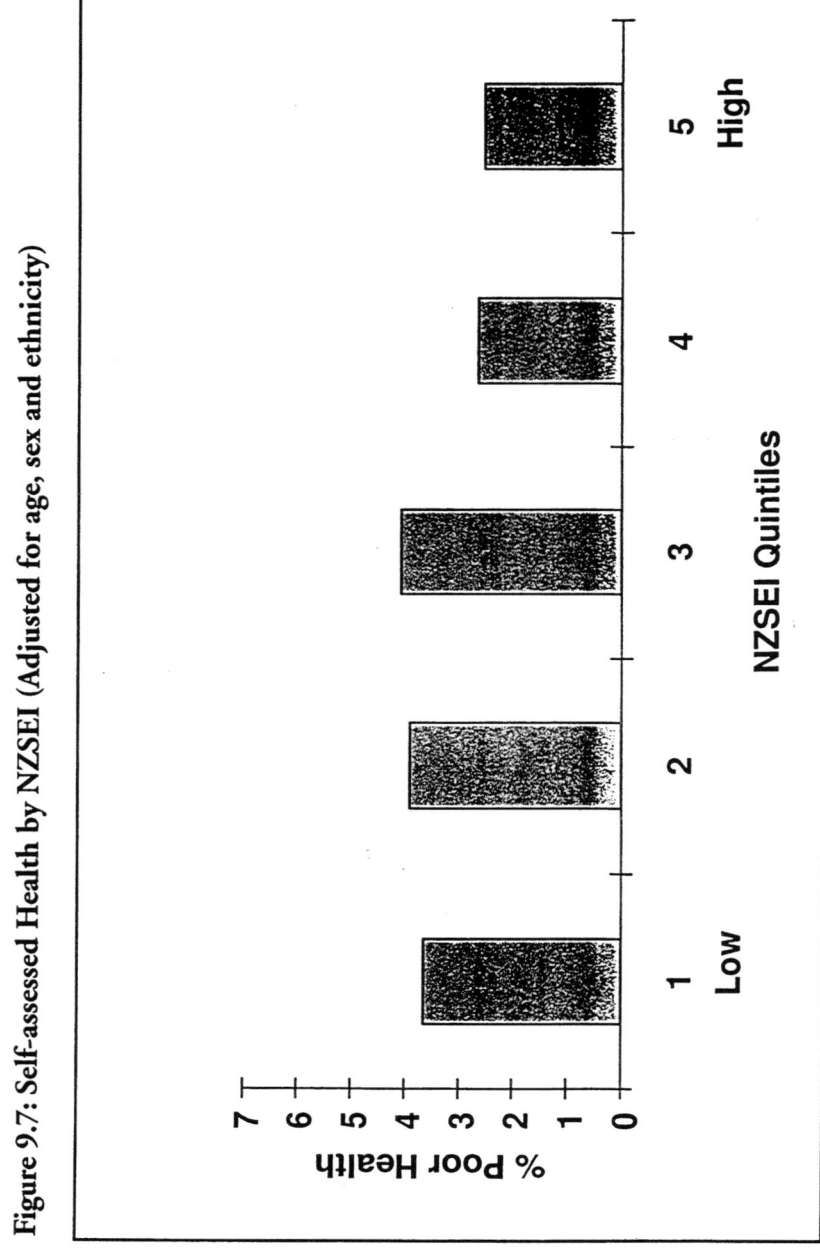

Figure 9.7: Self-assessed Health by NZSEI (Adjusted for age, sex and ethnicity)

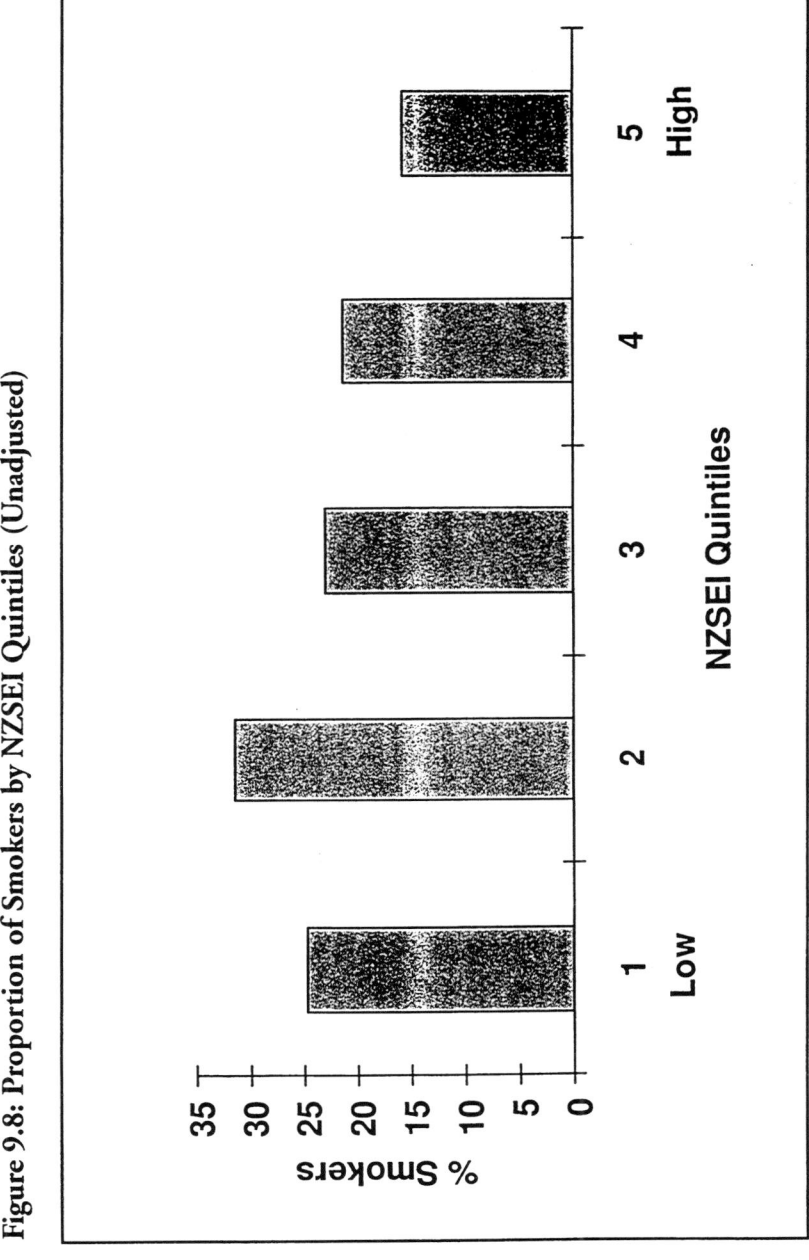

Figure 9.8: Proportion of Smokers by NZSEI Quintiles (Unadjusted)

quintiles. This result further contributes to the predictive value of the scale.

Discussion

The following observations can be made from our preliminary analysis. In the first place, the broad rankings of the occupational groups are consistent with expectation, except for the position of Agricultural and Fisheries workers (at the bottom of the hierarchy). Secondly, this consistency in ranking is maintained in the analysis of gender and ethnic differences. While males and Europeans have higher SES scores within each occupational group, the overall ranking remains much as it is for the entire data set. Nevertheless, the differences in scores are striking and provide an indication of the extent of gender and ethnic inequalities over and above those deriving uniquely from the occupational structure. Finally, the validation exercises using three health indicators are broadly consistent with expectation, with the notable exception of the bottom quintile of the NZSEI scale.

This last finding points to one problematic aspect of this and other SES scales – namely, the position of farmers and agricultural workers. In the context, two points should be made. Firstly, this category is very heterogeneous. In particular, it should be noted that the 1990 NZSCO revision amalgamated all agricultural workers, losing the distinction between farm owners and farm workers. Secondly, the income date from the Census does not accurately capture the overall economic position of farmers which is reflected more in asset base than necessarily in declared income. This can lead to questionable data accuracy and reliability. For example, Elley and Irving (1985) reported that the overall rating for farmers dropped from level 3 in 1976 to level 4 in 1981. In our published results for the NZSEI we will be treating agricultural workers as a special group.

Another area of interest that has already been alluded to is that of the striking differentials by gender and ethnic group. Why do women earn consistently less than men in almost all occupational groups? The explanation is more likely to relate to unequal opportunities than unequal education and ability. Moreover, in the case of some women, their current occupations may be part time or well below their capabilities because of the social expectation that they will assume the dominant role in child-rearing or taking care of elderly parents. Their SES may suggest a markedly lower overall position than these women might otherwise have or is

Figure 9.9: Proportion of Asthma Sufferers by NZSEI Quintiles (Unadjusted)

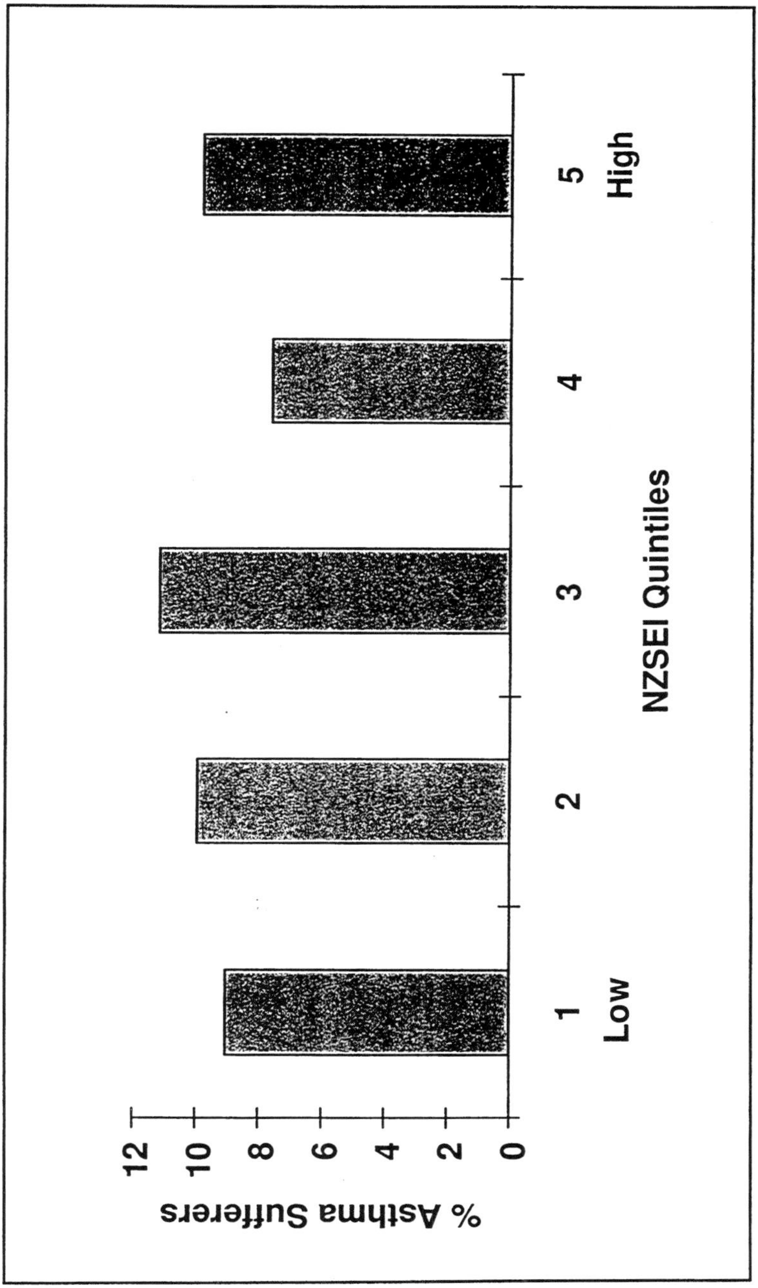

suggested, in the case of married women, by their husbands' occupations. In our published results we intend to use the NZSEI in a generic fashion for both men and women based on their current occupation. For our interim results, we have considered the woman's own occupation indicative of lifestyles and behaviour such as smoking, and those activities that are gender-specific, such as consulting the GP. However, in the long term, household socioeconomic characteristics are more important in determining overall access to scarce resources and thus life chances (Erikson, 1984).

Limiting social expectations create another related methodological issue in the pattern of occupations. Most women (Macran et al, 1994) and Maori (Brosnan, 1987) work in a relatively restricted range of occupations. Along with others, we have questioned whether previous scales developed on the full-time, largely male, mostly non-Maori, workforce can then be used to assess the impact of social class on the group lifestyles and life chances of women and Maori. The evidence from our Index shows that both women and Maori earn substantially less than we would expect for their education, age and occupations.

This indication of inequality is socially deeply concerning, but raises the methodological question as to whether using a single index enables us usefully to highlight these issues. In many ways the effect of ethnicity is conceptually distinct from the effect of occupational SES and it would be a mistake to adjust the Index for ethnicity. We have proceeded with a combined Index based on the proposition that on balance the same index can usefully incorporate both male and female, Maori and European, but we intend to do further exploratory work to try and explain the phenomena of their divergent incomes.

The position of the many adults who are not in the paid workforce at all, such as housewives, students, beneficiaries and the retired, is anomalous in an SES scale and highlights the limitations of an index of occupation to describe or explain SES. One approach is to use various occupational proxies, such as previous occupation or partner's occupation, but as well, we are exploring the feasibility of other markers of status. For example, Census data on the age-specific relationship between educational qualifications and occupational attainment for women could be used to impute a personal socioeconomic position for female beneficiaries and housewives, where this was consistent with our wider theoretical and practical purposes.

Conclusion

More than a decade ago, Jones and Cameron (1984) wrote a challenging article in which they lamented the banal categorical methods chosen to try explain the ubiquitous impact of social conditions on health. In the article they analysed the arbitrary basis of the first Registrar General's social class classifications and make a strong case that it, and subsequent revisions, were based predominantly on Stevenson's personal experiences, uninformed by any theory of society. When his first classifications did not smoothly predict rates of infant mortality, the outlying occupations were atheoretically juggled into different classes to make the associations more regular. This is an early example of data 'mining' – redefining the occupational variables better to fit the expected gradient.

Could our work in constructing an occupationally-based social class scale be called data 'mining'? We hope that this paper has convinced you otherwise. Unlike earlier scales, our Index is based on a plausible theoretical model, the returns to cultural capital centred on education; and to a reasonable extent predicts health behaviour that is strongly associated with the maintenance of health.

References

Brosnan, P (1987) 'Maori occupational segregation', *Australian and New Zealand Journal of Sociology*, 3, pp 89-103

Bunker, JP et al (eds) (1989) *Pathways to Health: The Role of Social Factors*, Menlo Park, California: Henry Kaiser Foundation

Duke, V and S Edgell (1987) 'The operationalisation of class in British sociology: theoretical and empirical considerations', *British Journal of Sociology*, 38, pp 445-63

Elley, WB and JC Irving (1972) 'A socio-economic index for New Zealand based on levels of education and income from the 1966 census', *New Zealand Journal of Educational Studies*, 7 (2), pp 153-67

Elley, WB and JC Irving (1976) 'Revised socio-economic index for New Zealand', *New Zealand Journal of Educational Studies*, 11 (1), pp 25-36

Elley, WB and JC Irving (1985) 'The Elley-Irving Socio-Economic Index 1981 census revision', *New Zealand Journal of Educational Studies*, 20 (2), pp 115-128

Erikson, R (1984) 'Social class of men, women and families', *Sociology*, 18, pp 500-14

Ganzeboom, HBG et al (1992) 'A standard international socio-economic index of occupational status', *Social Science Research*, 21, pp 1-56

Irving, JC and WB Elley (1977) 'A socio-economic index for the female labour force in New Zealand', *New Zealand Journal of Educational Studies*, 12 (2), pp 154-163

Jones, IG and D Cameron (1984) 'Social class analysis – an embarrassment to epidemiology', *Community Medicine*, 6 (1), pp 37-46

Macran, S et al (1994) 'Women's socio-economic status and self-assessed health: identifying disadvantaged groups', *Sociology of Health and Illness*, 16, pp 182-208

Najman, JM and M Bampton (1991) 'An ASCO based occupational status hierarchy for Australia: a research note', *Australian and New Zealand Journal of Sociology*, 27, pp 218-213

Public Health Commission (1993) *Our Future Our Health. Hauora Pakari, Koiora Roa. The State of Public Health in New Zealand 1993*, Wellington: PHC

Statistics New Zealand and Ministry of Health (1993) *A Picture of Health*, Wellington: Statistics New Zealand and Ministry of Health

10 ~ The *NZDep91* Index of Deprivation

Peter Crampton, Clare Salmond and Frances Sutton

Introduction

This paper discusses a new index of deprivation for use in New Zealand. The index, called *NZDep91*, measures deprivation at a population level. Its role is complementary to that of individually-based classifications of occupational class.

The *NZDep91* index has been developed with three main purposes in mind: as a tool for resource allocation, as a tool for research and for use in advocacy at a community level.

Rationale for Developing a New Index

In September 1994 the Health Services Research Centre convened a meeting to discuss measures of socioeconomic status. The aims of the meeting were to find out which socioeconomic status indicators government agencies and researchers were using, and to discuss the adequacy of these indicators. It was clear that many sectors, including health, education, police and the fire service, were potential users of measures of socioeconomic status for the purpose of targeting funding to services.

The meeting concluded that there was demand for a set of standardised socioeconomic status indicators which could have application across a range of sectors. Further, it was clear, especially in the context of funding primary health care services, that socioeconomic status needed to be measured at the smallest possible population level.

There were other justifications for this research. Firstly, there was a decade of research to build on since Reinken et al did their original work in New Zealand. Secondly, and importantly, the notion of targeting resources for social services is now widely accepted, and has bipartisan political support. In order for targeting policies to be properly implemented, increasingly accurate measures of socioeconomic status and deprivation will be required.

Definitions

The key theoretical concept underlying this research is deprivation. Deprivation has been defined and described by Townsend (1979) in the UK as:

> ... a state of observable and demonstrable disadvantage relative to the local community or the wider society or nation to which an individual, family or group belongs.

It is important to note that deprivation, as defined, is a relative state rather than an absolute state. Townsend distinguished material from social deprivation. Material deprivation refers to material apparatus, goods, services, resources, amenities and physical environment and location of life. Social deprivation refers to the roles, relationships, functions, customs, rights and responsibilities of membership of society and its subgroups. The *NZDep91* index of deprivation reflects aspects of both material and social deprivation.

Deprivation refers to actual disadvantage in terms of social and material resources. Therefore, ethnicity and age per se are not included in *NZDep91*, as they are not direct measures of either material or social deprivation, and cannot be directly influenced by social policy.

Poverty and socioeconomic status can be distinguished from deprivation. Poverty has been defined as lack of the resources necessary to avoid material and social deprivation. The New Zealand Royal Commission on Social Security (1972) took a relative approach in defining poverty somewhat similar to Townsend's approach to deprivation:

> The problem of poverty in developed industrial societies is increasingly viewed not as sheer lack of essentials to sustain life, but as an insufficient access to certain goods, services and conditions of life which are available to everyone else and have come to be accepted as basic to a decent minimum standard of living.

The 1988 Royal Commission report focused more on participation, stating that the objectives of income redistribution policies include the need:

> To ensure that all New Zealanders have access to a sufficient share of income and other resources to allow them to par-

ticipate in society with genuine opportunity to achieve their potential and to live lives that they find fulfilling. In so doing to provide a measure of certainty and security for all throughout their lives.
To relieve immediate need arising through unforeseen circumstances.
To ensure the well-being and healthy development of all children.

Socioeconomic status is a concept related to deprivation and poverty, encompassing both poverty and affluence. Socioeconomic status is a descriptive term for a person's position in society, which may be expressed in terms of income, education, occupation, value of dwelling place and so on.

Existing Indices of Deprivation

Health and equity scores (HEQ scores) were developed by Reinken et al in the early 1980s. HEQ scores measure deprivation at the level of census area units, which commonly hold about 2,000 people. HEQ scores were subsequently used in population funding formulae for area health boards, and later for regional health authorities. HEQ scores have also been used in research and advocacy.

The only other New Zealand index of deprivation that we are aware of in the health sector is the Midland Index of Relative Disadvantage, which was developed by Midland Health. This index has been used in needs assessment in the Midland region. Across the Tasman the Australian Bureau of Statistics has developed a set of five composite indices, including indices of deprivation. By far the most work, however, in the general area of indices of deprivation has been carried out in the UK, where a number of indices have been used for research and resource allocation since the early 1980s. UK indices include the Townsend index, the Jarman index and the Scottish Deprivation index (there are several others).

The *NZDep91* Index

The index described here is the best of sixteen that have been explored in this study.

There are ten variables in the index. All variables were constructed from questions in the 1991 Census conducted by Statistics New Zealand.

Each variable is the proportion of persons in an area with a lack of something – income, access to personal transport, living space, an owned home, employment, qualifications or personal and/or financial support.

Two of the variables – household income and occupancy – have been adjusted to reflect family size and structure. Household income has been adjusted by use of the Jensen scale. Living space, or occupancy, is the ratio of the number of adult equivalents in the household to the number of bedrooms, where each child under ten counts as half an adult equivalent. Where a variable reflects a household characteristic – household income, occupancy, car ownership and home ownership – each person in the household is ascribed that characteristic for the purposes of calculating proportions.

Specifically, the ten variables are proportions in an area:

- with equivalised household income below a threshold (set to include 15% of households);
- with a means-tested benefit and aged 18-59 years;
- without access to a car;
- with an equivalised household occupancy ratio above 1;
- not living in a household which owns the dwelling (home);
- unemployed and aged 18-59 years;
- without qualifications and aged 18-59 years;
- in a single parent family;
- separated or divorced and aged 18-59 years;
- separated or divorced and aged 60 years or over.

Since relative deprivation for an area will depend, to some extent, on the demographic make-up of that area, each variable has been indirectly standardised for the age and gender distribution in each area. Parallel indices with unstandardised variables have also been created. Comparison of standardised and unstandardised indices suggests that standardisation is necessary.

The building blocks for the small areas are meshblocks, the smallest aggregation used by Statistics New Zealand for census purposes. There are about 35,000 meshblocks, having a median size of about 90 persons. Clearly this implies that some meshblocks would be far too small for calculation of meaningful and robust proportions.

The primary sampling units used by Statistics New Zealand for their household surveys have been used to agglomerate these meshblocks into somewhat larger, contiguous areas. Primary sampling units contain one or more meshblocks and at least 30 households. Some primary sampling units contain several hundred persons.

Two types of agglomeration of meshblocks were explored: to populations of at least 100 persons, and to populations of at least 200 persons. The agglomeration procedure follows a simple computerised algorithm to form the small areas. Essentially, large primary sampling units are broken into component meshblocks, or groups of meshblocks, with at least 100 (or 200) persons in each, while smaller primary sampling units remain intact. For the smaller aggregation, 95% of the resulting 20,154 small areas have 100 or more persons usually resident in them, and only 1% have a population less than 80. The minimum population is about 40.

A standard principal components analysis was used to create an index from the ten deprivation variables, the index being the first principal component and thus a weighted sum of the proportions – of each of the ten variables – in any particular small area. The index scores are scaled to have mean zero and standard deviation one although the distribution is somewhat skewed with a minimum of -1.9, a median of -0.2, and a maximum (most deprived area) of 5.3, and with 1% of values greater than 3.0. A ten-point ordinal scale for deprivation has also been created, with values 1 (least deprived area) to 10 (most deprived area) being defined on the basis of percentiles. The cut-points occur at 10, 20, 30, 70, 80, 85, 90, 95 and 97.5% of the distribution respectively.

'Validation' of the index in the absence of a gold standard consisted of investigations of technical aspects of the index, exploration of scores in sentinel areas and correlation of the index with key health variables.

Technically, the variables make sense and their weights are consistent with expectations. There are no large pockets of missing data, in particular in the smaller of the small areas. The weights for aggregations to 100 and 200 persons are very similar, suggesting that the 100-level aggregation can safely be used unless pockets of instability are found.

Deprivation scores in sentinel areas chosen on the basis of local knowledge, in particular in Taita and other areas around Wellington, were as expected. None of the small areas with the smallest populations have unlikely deprivation scores. There is no unexpected pattern to the demography for the areas of greatest deprivation: the 41 small areas with principal component scores greater than 4.0 included both urban and rural areas from many parts of New Zealand, and only four had populations fewer than 100.

The final step in the validation process was to correlate the index with certain health outcomes well known to be associated with deprivation. Three data sets were explored: mortality from all causes in the Wellington region in the period 1990-1993, hospital discharge ratios for the same region and period, and national registrations for lung cancer for the same period. Each data set was explored in four broad age bands to remove most of the confounding effects of age. The picture from these analyses is quite clear: areas of increased deprivation experienced increased mortality rates, increased hospital discharge ratios and increased registrations for lung cancer.

Conclusions

We would like to stress that work on the index is not complete. Although *NZDep91* appears to be a sound index of deprivation in small areas, comments and further confirmation will be sought from selected users of a documented beta-test version of the index which will be made available in February/March 1997.

Those interested in having a beta-test version early in 1997 should contact Clare Salmond or Peter Crampton at the Health Services Research Centre, or the Department of Public Health, Wellington School of Medicine.

Ethics and Confidentiality

Ethical approval was obtained for this research project from the Central Regional Health Authority Wellington Ethics Committee.

To protect against disclosure of information supplied by individual respondents, Statistics New Zealand's practice is to round all aggregated Census output randomly. In this case access was granted to unrounded aggregate Census data, under a special contract between Statistics New Zealand and Clare Salmond and Peter Crampton, so that the scarcely populated meshblocks could be sensibly agglomerated to form larger areas. The access was granted in a strictly protected environment on Statistics New Zealand premises, under supervision of Statistics New Zealand staff, and no unrounded census aggregations were removed from the site.

Both researchers are bound by the same provisions of the Statistics Act 1975, which bind staff of Statistics New Zealand to preserve the confidentiality of individual respondent data.

Acknowledgements

The project was funded by the Health Research Council, with financial support also from Statistics New Zealand. Equally important has been the tremendous personal support which Statistics New Zealand has provided, in particular Len Cook, Mark Strang, Mike Brookes and Lyndsay Gillespie.

References

Carstairs, V and R Morris (1989) 'Deprivation: explaining differences in mortality between Scotland and England and Wales', *British Medical Journal,* 299, pp 886-889

Castles, I (1994) *Information Paper, 1991 Census Socio-Economic Indexes for Areas,* Canberra: Australian Bureau of Statistics

Crampton, P and M Laugesen (1995) *The Use of Indices of Need in Resource Allocation Formulae for Primary Health Care,* Discussion Paper No. 3, Wellington: Health Services Research Centre

Jarman, B (1983) 'Identification of underprivileged areas', *British Medical Journal,* 286, pp 1706-1708

Jensen, J (1978) *Minimum Income Levels and Income Equivalence Scales,* Wellington: Department of Social Welfare.

Jensen, J (1988) *Income Equivalences and the Selection of Family Expenditures on Children,* Wellington: Department of Social Welfare

Kokaua, J (1993) *Computation and Validation of Midland Health's Index of Relative Disadvantage* (unpublished paper), Hamilton: Midland Health

Last, J (ed.) (1995) *A Dictionary of Epidemiology,* Third Edition, New York: Oxford University Press

McLennan, W (1990) *Socio-economic Indices for Areas,* Catalogue no. 1356.0, Canberra: Australian Bureau of Statistics

Morris, R and V Carstairs (1991) 'Which deprivation? A comparison of selected deprivation indices', *Journal of Public Health Medicine,* 13, pp 318-326.

Reinken, J et al (1985) *Health and Equity,* Wellington: Department of Health

Royal Commission on Social Security (1972) *Social Security in New Zealand*, Wellington: Royal Commission on Social Security

Royal Commission on Social Policy (1988) *The April Report, Social Perspectives*, Wellington: Royal Commission on Social Policy

Townsend, P (1987) 'Deprivation', *Journal of Social Policy*, 16, pp 125-146.

Urlich Cloher, D and L Murphy (1994) 'Maori housing needs (Northland)', Proceedings of the Housing Research Conference

11 ~ Workshop Reports

Report from Research Priorities Workshop
Facilitated by George Salmond, reported by Sharron Bowers

Table 11.1: Key Areas of Discussion by the Research Priorities Stream

• Methodological issues in research into health impacts of socio-economic inequalities
• Priority areas for socioeconomic inequalities and health research
• Collaborative opportunities for research into socioeconomic inequality and health

Methodological Issues in Research
Developing Frameworks and Tools
It is important to develop our measurement tools and models for explaining interactions between SES inequalities and associated health outcomes. However, we must not lose sight of why we are doing the research. Endless debate about sophisticated measures will not reduce health inequalities. The aim must always be effective public health action.

We need to focus more attention on information and frameworks which show the impact of socioeconomic status on health

There is a need to go beyond mortality data in measuring SES inequalities and health. We often mean mortality when we talk about health measures. Mortality is the lowest common denominator. These measures alone are inadequate to show the full impact of SES inequality on health.

Longitudinal data is lacking for many health measures. One-off measures are useful but we need continuous data to show trends.

More small area approaches are required to show the distribution of inequality. Aggregate measures hide regional and local differences and are not that helpful in planning services and allocating resources at the local level.

Linking Data Sets
One possibility for overcoming gaps in data is to link data sets.

157

For example, the Canadian National Population Health Study asks participants if they can have access to their health insurance data. In this way it is possible to link data gained from each individual in the health survey, with their health history as collected by the insurance company data.

Poverty measurement work focuses on housing and income measures and seems to separate poverty measurements from health measurements. More attention needs to be paid to correlations between the two data sources. One possibility for linking measurement of deprivations is to link poverty line data with the new deprivation index *NZDep91* which has recently been developed by the Health Services Research Centre.

The practice of rounding off data is a barrier to linking of data sets. Statistics New Zealand rounds off data to ensure that individuals cannot be identified from statistical information. Unrounded data is difficult to obtain from Statistics New Zealand which works under strict security and provisions of the Privacy Act. Statistics New Zealand is presently working with researchers to address this issue.

Another barrier to linking data sets is public concern that linking data sets will make personal information available in an unacceptable way. If public confidence is not to be undermined and research jeopardised, researchers must take their data security responsibilities very seriously.

Determining What Gets Measured

Compared to other countries, New Zealand has only a small pool of resources for data collection. We need to be selective about what information we collect.

Traditionally, New Zealand has made only very limited investments in R & D and information services. Currently, there is no health sector funding set aside nationally to support these services. R & D and information services are not linked to a coordinated and public process for setting strategic direction for the health sector as a whole. As a result, information gathering and the R & D effort are not targeted on priority problems such as poverty and health. This makes it difficult for researchers to obtain the research resources they need, including information.

New Zealand lacks a coordinated approach from research communities in choosing what statistics are required. For this reason, Statistics New Zealand sometimes has difficulty in identifying users' needs. A coordinated approach would be welcomed.

Multi-method Approaches
Research efforts need to encompass multiple methods including qualitative and quantitative, subjective and objective, local and global, high risk and population-based approaches. We need to juggle as many of these balls as possible, without falling into binary views that one type of information is 'better' than another. Multimethod approaches require a collaborative approach to research.

The HRC Funding Process
The current HRC funding process has major limitations. In competition with established bio-medical researchers, it is difficult to get resources to initiate research in new and difficult areas.

The process is time consuming and somewhat unpredictable. It favours relatively straightforward single discipline research rather than collaborative multidisciplinary research. And it does not take account of the time and effort required to build research capacity in areas such as poverty and health.

Steps are urgently needed to put in place, in New Zealand, a national R & D strategy in the health sector similar to that currently operating successfully in the United Kingdom. Within such a strategy 'poverty and health' research should be given high priority.

Research Priority Areas
Workshop participants identified the following needs for research. The list does not indicate relative priority.

Literature Review
There is a rapidly growing body of research overseas concerning relationships between socioeconomic inequalities and health. It is essential that researchers in New Zealand keep up to date with developments internationally. A reference point should be mandated to ensure that relevant information is readily available to the interested research community in New Zealand.

Establish a Poverty Line
There is clearly a need to consolidate the poverty line research currently being undertaken by Charles Waldegrave and colleagues. At some point a methodology should be established and resources made available to establish an official ongoing measure of poverty in New Zealand. We will get

more mileage out of poverty research with an accepted poverty line against which to show the impacts of changes in social and economic policies.

Once a poverty line is established, research is needed to demonstrate linkages between poverty and poor health. To this end, the work being done at the Health Services Research Centre on the new deprivation index is a step in the right direction.

Show the Costs of Not Dealing with Poverty

Research shows that socioeconomic inequality is a significant cause of illhealth. A major issue is getting this understood by the public and by government. Policy-makers need numbers attached to social problems before the impact of social and economic policies becomes 'visible'. Measurement must be improved.

International comparison is a powerful tool. Seeing New Zealand relative to other countries we consider to be our equivalents, such as Canada and Australia, can have a powerful effect on policy. Investing in an international study like the Luxembourg study on income and expenditure would place New Zealand in international league tables, providing a ranking for New Zealand relative to other countries. The Luxembourg survey compares income patterns of participating countries. Some countries also contribute expenditure data. New Zealand census does not currently satisfy Luxembourg requirements. The direct cost of New Zealand participation in the Luxembourg survey could be as little as $10,000 per year.

Similarly, comparison of variation within New Zealand is a powerful tool for showing the impact of SES inequality on health

Make Explicit the Impact of Values in Service Provision

Research is required which looks at how health services and health providers respond to poverty.

Attitudes and values of providers can influence the quantity and quality of health care received. Providers may knowingly or unknowingly exclude people from services and information or treat people differentially on the basis of ethnicity and socioeconomic status.

Research in this area could make these barriers explicit. For example, research done by John McKinley involved sending actors with the same case history to primary health care practitioners. He found that wide-ranging treatments were prescribed for the same 'condition' based on ethnicity or the socioeconomic status of the 'patient'.

Examine the Intervening Factors between Poverty and Illhealth
While the link between poverty and health status is well established, intervening factors are not so well understood. Studies using a variety of research approaches are needed to tease out these factors. Many factors could be involved. For instance, widening income differentials may be reducing the capacity for relatively poor people to participate in society.

Inability to participate may alienate significant sections of the community leading to reduced social cohesion. Lack of social support may be conducive to damaging personal behaviours such as tobacco use, drug abuse and domestic violence. Longitudinal studies are needed to explore possible linkages and chains of causation.

Show Impact of Community Cohesion/'Social Capital' on Health
Community cohesion, which is based on levels of trust and goodwill in society, is related to the health of communities. Research which shows the impact of social and economic policies on community cohesion and its relationship with health would be valuable.

The researchers who introduced the idea of 'social capital' did so to relate the concept to the dominant economic paradigm and attract the interest of economists. To this end they have certainly been successful. But capital accumulation is a western cultural value and is not appropriate to all cultures. The relevance of metaphors used to describe this concept could be examined and other metaphors explored.

New Zealand has, in the past, prided itself on the extent and vigour of its voluntary organisations. These have undoubtably contributed to the social cohesion of New Zealand society. But what is happening to these organisations and their associated activities today? How is our society changing in this regard? And what is the health impact of these changes?

There are a number of sources of information which currently provide pieces for this jigsaw puzzle. Examples include Citizens Advice Bureaux client statistics which give an indication of the role that various voluntary organisations play in providing services to those who fall between the gaps. Many Maori initiatives specifically aim to build social solidarity and thus promote health. These should be evaluated.

Raise the Voices of Those Affected by Socioeconomic Inequality
Advocacy is required to raise the voices of people affected. This needs to be done at both public and policy levels.

The process of gathering and sharing information should be an empowering process for participants and communities. Research has value not only as an exercise in gathering information, but also in building social capital and in informing and encouraging 'voice'.

Research needs to be sensitive to the stigmatising effect of labelling communities and individuals as deprived. Research should done 'with' not 'on' communities.

Participatory research is not only ethical but practical. It is increasingly difficult to get communities to work with researchers because of previously badly handled or misunderstood research efforts. Research which is participatory and empowering works.

Developing social cohesion between researchers and communities affected by inequalities is an important step to raise the profile and understanding of poverty and its relationship to health.

Collaborative Opportunities for Research
Develop Framework of Association
There is currently no structure or framework which brings people together to collaborate on socioeconomic inequality and health research. The presently highly competitive environment discourages collaboration.

The framework for setting and reviewing 'Strategic Directions' established by the Public Health Group in the Ministry of Health provides an overriding framework for the advancement of public health. But a more precisely targeted approach is required especially in areas such as poverty and health. Part of that targeting should include resources to promote a well-orchestrated and collaborative research effort to address key questions pertaining to poverty and health. To be really effective this research should be managed, not ad hoc, and should involve all interested parties, not just the academic community.

The workshop was encouraged by the steps the government is presently taking towards development of a national R & D strategy for the health sector. Progress currently being made towards the targeting of research in support of national mental health strategy was applauded. Similar steps should be taken in support of public health strategy, particularly in the area of poverty and health.

Existence of a Research and Development fund would provide a 'space' for national collaborative projects and ensure that all findings were available in the public domain.

Acknowledge Process
Any collaborative effort will bring together people from diverse backgrounds who have different perceptions of process, despite having shared goals. Tensions will naturally arise and need to be accepted and handled, working out ways of creating synergy from diversity rather than being pulled off course by it.

Research issues need to be clarified at the outset to ensure that mutual interests are met and differences resolved.

People affected by socioeconomic inequality should be actively involved in the process.

Shared vision is vital to the success of collaborative efforts.

Summary

The workshop on research priorities covered three areas: methodological issues, priority areas, and collaborative opportunities for research into socioeconomic inequality and health.

Methodological issues raised included the need to develop frameworks and tools, determining what statistics get collected at a national level, making best use of data by linking of datasets, using multiple methods to ensure all sides of the 'story' are seen, and limitations of the HRC funding process.

Research priority areas identified included the need to keep up with international literature concerning socioeconomic inequalities and health, establishing an official poverty line for New Zealand, raising understanding of the causes of poverty, showing the costs of not dealing with poverty, making explicit the impact of values in service provision, examining the intervening factors between poverty and ill health, examining the relationship between community cohesion and health, and raising the voices of those affected by socioeconomic and health inequality.

Collaborative research opportunities require a framework of association and acknowledgment of inherent process issues.

Report from Purchasing Stream Workshop

Facilitated by Fran McGrath, reported by Mavis Duncanson

For purchasers of health services, indices of deprivation are important to enable funding of health services in a way that reflects need and that is likely to achieve equitable outcomes. There was consensus about the importance of reallocating resources according to health need, and acceptance of deprivation as an indicator of health need.

Table 11.2 summarises the issues raised and debated in the wide-ranging discussion of this group.

Table 11.2: Key Areas of Discussion by Purchasing Stream

• Use limitations of current tools for measuring deprivation and targeting resources
• Barriers to equitable allocation of resources
• Need for intersectoral collaboration and cooperation
• Identified areas of good practice

Current Measurement Tools

1. Community Services Card

The community services card is designed to reallocate health resources to target individuals, identified in relation to family income and composition. It does not take account of need attributable to the following factors:

- ethnicity
- age
- gender
- availability of childcare
- living in a rural area
- affordability of services
- cultural factors.

Access to health services is restricted for family members, including financially dependent young people, who do not have access to the family's community services card.

164

The card is also unsuitable for members of itinerant populations, arguably in need of targeted assistance, due to the administrative difficulty in keeping records updated (not to mention keeping track of the card itself).

The targeting of families to receive assistance cuts across social connectedness, creating unnecessary and artificial distinctions between people in a community.

Alternative Proposal
An alternative to the principle of targeting is the provision of universal primary health care with differential funding according to the level of need in the community.

2. Health and Equity Score (HEQ)
RHAs have used HEQ scores in locality profiling to provide a general indication of need in an area. The size of the units used obscures variation within an area.

3. NZ Deprivation Index
Identified advantages of this new index include:

- the ability to measure deprivation at meshblock level enhances accuracy of the measurement for any given locality;
- provision for the possibility of differential funding without the disadvantages of targeting individuals and families;
- the use within a capitated service of routinely collected information (such as addresses of service users) to provide ready access to a measure of need in the practice group. This process requires minimal additional administrative time.

Participants noted that the index cannot quantify need associated with other variables within a community such as the incidence of sexual abuse, or the use of alcohol and drugs.

4. Community Consultation
Tools to measure need require augmentation with "connectedness with the community". Community consultation processes provide the opportunity for members of a community to articulate the realities of life in the

community. Factors other than cost may restrict access in a particular community. For example, parents of young children may avoid or delay attending to their own health needs because of the cost of childcare, or the inconvenience of taking children to a doctor's surgery. The level of unemployment, or the provision of public transport may affect the accessibility of services.

Participants noted that the obligation to consult with local communities requires and empowers RHAs to access and use qualitative information in making purchasing decisions.

Equitable Allocation of Resources

There was agreement that the application of the population-based funding formula had resulted in more equitable allocation of health resources at the RHA level. However, equitable allocation of resources had not been achieved below this level. Barriers to equitable allocation of resources include:

- a lack of political will at all levels to reallocate resources to people who are implicitly blamed for their status in the community;
- a lack of local political will in communities that stand to lose resources in the reallocation process. The reality of loss of employment, and subsequent effects on local economies, is a significant factor for local communities;
- the competitive environment among providers of health services which inhibits sharing of information and rational decision-making;
- the focus of purchasing on services or outputs rather than on health outcomes;
- funding of independent practitioner associations on an historical basis which allows those in affluent areas to make savings and provide even more services to their practice populations. Conversely, those who improve access for low income populations face expanding costs with less opportunity to provide additional preventive services. A challenge would be to use deprivation measures, rather than historical spending, as the basis for budget setting;
- media response to attempts to reallocate resources tends to consolidate public opinion (arguably middle class public opinion) against change and does not necessarily reflect the view of the community as

a whole. Those who continue to miss out on services are often the least articulate members of the community who do not have a political voice. The self-determined priority of public health initiatives for these groups is seldom if ever picked up by the media, and is overshadowed by more vocal public demands for increasing secondary and tertiary services;

- expectations of the health system vary in relation to advantage and disadvantage, so that those who have least also expect least. Low self-esteem, social isolation, an inability to participate in the economy, and poor information were seen to contribute to this imbalance of expectations. This dilemma was expressed as "taking from the rich, who then feel deprived, to give to the poor who don't realise they could have more". Those in most need have less voice *and* less awareness of their need than those for whom there is better provision;

- the health reform process was instituted within a short time frame, with inadequate attention to the management of change. The purchaser/provider split has not improved access to health services, nor improved the health of the most disadvantaged people in our communities;

- there has been inadequate transitional funding to maintain existing services while new patterns of service delivery are developed and introduced. The comment was made that it takes time and money to build up the necessary infrastructure to provide a quality service. This does not happen overnight nor does it happen on the smell of an oily rag. It very often happens as a result of the unpaid and unrecognised work of women in a community;

- there is inadequate information about how services contribute to health gains to allow marginal analysis and shifting of resources within programme budgets (Programme Budgeting Marginal Analysis or PBMA);

- while there appears to be political awareness of the connection between poverty and illhealth, this awareness is not always apparent among boards and management of RHAs. There was a perceived need for a 'roadshow' to disseminate the findings of this conference to those responsible for developing purchasing policy.

Participants noted that need is not the only basis for allocation of resources. Purchasing policy would be clarified by exploration of the relationships and conflicts between the following:

- indices of deprivation;
- priority health gain areas; and
- Programme Budgeting Marginal Analysis.

Integrated Purchasing of Services
There was a clearly expressed need for intersectoral collaboration and co-operation in the development of policy and service provision. The following observations illustrate this need:

- providers are often required to report to several agencies, yet there is no allowance for these additional administrative costs in their contracts. The challenge to health providers is to state that fundamental structural change is necessary to improve the health of communities;
- concern to eliminate 'double dipping' has created gaps in the availability of services;
- within Vote: Health the ringfencing of public health, mental health and disability support services has meant that some services fall outside all fences;
- one participant indicated that there are 24 government agencies that have an interest in health.

Participants presented a challenge for health service providers to lobby for structural change which recognises this reality and allows for a coordinated approach.

Participants noted that there is already considerable intersectoral collaboration, particularly at ministry level. However, the links have to be established at all levels, including the local community. These levels can be conceptualised as:

- Policy
- Purchasing
- Providing
- Participating.

Community involvement is essential to develop appropriate services. This was expressed as the need to develop services to fit client groups, rather than expecting client groups to fit the services provided.

From a consumer perspective there is a need for integrated service provision with a clearly defined point-of-first-contact. One participant expressed this as consumers' need for a "one stop shop". In purchasing services for particular groups in the community, integration between health policy and labour market, education, environmental and other policy areas is important.

The fundamental question of whether it is possible to 'purchase' social cohesion was left unanswered.

Examples of Effective Collaboration
The following examples of successful collaboration and cooperation were raised during discussion:

*   interdepartmental and interagency consultations do result in joint planning and policy development. For some participants there is not a lot of evidence of this at a local level;
*   funding agencies have worked together to streamline their reporting requirements;
*   North Health has produced and distributed to residents of the region a 'Service Guide' to improve access to information about available services. An 0800 telephone number is provided for queries. A certain degree of literacy and access to a telephone are necessary to benefit significantly from this initiative;
*   the case management system implemented by Income Support has improved coordination of services provided to many clients; and
*   a district council initiative in Waitakere has led to improved integration of services at a local level.

These examples show what is possible. However, participants recognised a need for much greater structural change. In relation to the conference, intersectoral collaboration on the use of tools for measuring deprivation and allocating resources would be useful.

Political will to reallocate resources on the basis of need is a fundamental requirement. Participants recommended issuing a request for proposals to the politicians in the new Parliament.

Request for Proposals
Proposals are called for to provide a more equal society, in order to improve the health status of all New Zealanders.

Report from Maori Stream Workshop

Facilitated by Paora Howe, reported by Clive Aspin

Te hauora o nga iwi Maori
ko nga korerorero o nga tangata o te Papamahi Maori

Socioeconomic Inequalities and Maori Health

Kia mihia te mano tini kua mene ki nga Hawaiki katoa,
ratou te tututanga o te puehu,
te whiunga o te kupu i nga wa takatu ai ratou.
Heoi, waiho ake ratou ki a ratou,
tatou te urupa o ratou ma,
nga waihotanga mai e hapai nei i o ratou wawata,
tumanako hoki.
Kia ora tatou

The Maori stream of the Socioeconomic Inequalities and Health Conference saw Maori and Pakeha from diverse backgrounds come together to discuss issues related to the health of Maori both in the present and in the future. Community workers, researchers, policy analysts and health professionals identified areas and issues that require urgent attention if the health status of Maori is to be addressed and improved. Discussion was based on a profound understanding of the practical and political realities of Maori health in the 1990s. It was a discussion that was driven by a strong desire to see an improvement in Maori health in our lifetime so that future generations might benefit from the directions and initiatives that have been implemented within the last decade.

Participants identified a number of key priorities that need urgent attention if we are to see any visible impact on Maori health:

- data collection, health measures and analysis;
- resource allocation and funding;
- workforce development;
- research priorities and links with the policy process;
- community development;
- housing; and
- Maori control over decision-making.

170

Data Collection, Analysis and Health Measures

For too long decisions about Maori health have been based on information that gives an inaccurate understanding of Maori health and the factors that impact on Maori health. Practitioners in the workshop cited examples of misclassification of data related to Maori and the effect that this has on the delivery of services to Maori. Several participants reiterated the need for the implementation of mechanisms that would lead to the collection of accurate data relating to Maori health. Currently, the collection of ethnicity data of hospital admissions is poorly done and this has serious implications in the areas of resource allocation, access to services and the funding of Maori health initiatives. There must be a commitment from decision-makers to rectifying this anomaly as a foundation to the improvement of the health status of Maori.

Accurate data collection must be coupled with appropriate analysis and both must be underpinned by appropriate measures of Maori health. This triangle will allow for a more accurate appraisal of Maori health than that which is currently in place.

There must also be a commitment to the development and implementation of appropriate means of measuring Maori health. The instruments must be culturally appropriate. There was wide support for the model presented by Crampton, Salmond and Sutton. Participants felt that there was an urgent need for the development of a poverty index. Such an instrument would give a more accurate picture of the overall health status of New Zealanders in general and Maori in particular.

Resource Allocation and Funding

Accurate methods of measuring and analysing Maori health will allow for fair and equitable allocation of resources. Participants referred to instances where this is not the case and stressed the need for this to be addressed urgently.

Reference was made to the difficulties that often confront Maori health initiatives. Constraints related to inadequate funding and short-term contracts mean that these initiatives are often prevented from carrying out their work in a way that will have beneficial impacts on Maori clients. Examples were cited of the contractual difficulties that confront Maori health providers in their dealings with regional health authorities.

The Development of Research Priorities and Links with the Policy Process

There is an urgent need for the development of Maori health research priorities. These must inevitably come from the community which must be given the opportunity and resources to identify areas of priority. Once they are identified, these research priorities must then be actioned by appropriate research bodies.

The role of the Health Research Council was discussed at some length. The council has a crucial role to play in all areas related to Maori health research. It is vital that the HRC continues to fund Maori initiatives at an adequate and appropriate level, and it must help to ensure that community concerns and priorities are articulated and heard. Adequate funding for community initiatives will facilitate the development of appropriate Maori health research.

Research is a vital component of sound and effective policy. For Maori health policy to be effective it must be based on robust research that has been developed by Maori with significant input from those who are most affected by the research process and outcomes.

At present, the links between the research process and the policy process are either non-existent or ill-defined. For research to have desired results it must be closely tied into the policy process. Ways of developing and consolidating these links must be explored.

Several participants felt that more transparent means of communication were vital to the establishment of effective links between not only researchers and policy-makers but also among policy-makers. At times within both arenas there tends to be duplication of work which could be avoided if channels of communication were improved.

Some participants called for regular Maori health hui. A planned meeting schedule would provide Maori health workers with a venue so that they could meet on a regular basis and explore issues of concern. The establishment of a formalised network is integral to developing effective methods of communication.

Workforce Development

For some time Maori have been disadvantaged by the lack of skilled Maori health workers. Funders of Maori health initiatives have a key role to play in facilitating the development of adequate numbers of skilled workers. This applies to all areas of Maori health.

Participants acknowledged that the HRC has developed a strategy for the development of the Maori health workforce. However, caution must be exercised over how we participate in this strategy. Currently, there is a dearth of information on Maori health and this means that it is difficult to make accurate forecasts. Over time, the development and training of Maori health researchers will help to rectify some of the problems that result from the misclassification of Maori health information.

Community Development

Community development is fundamental to improving the health of Maori. Funding must be made available to ensure that this happens. Furthermore, it is important to identify the links between workforce development and community developments so that efforts are not duplicated.

Community workers at the workshop stated their desire to see communities being given the means whereby they are able to make decisions for themselves. Researchers and policy-makers have an important complementary role to play in ensuring that communities are empowered to advocate for change and improvement.

There was recognition that some communities were more deprived than others. Such communities need to be identified and then there must be a willingness to move resources to the areas of greatest need.

Housing

The link between poor housing and poor health was discussed at length. For many Maori the high costs of adequate housing are prohibitive and this has a direct impact on their health status. There must be an indication from the relevant policy-makers that this is an area of crucial concern. Research into issues related to housing was seen as vitally important to improving Maori health status.

Maori Control over Decision-making

Overwhelmingly, participants reiterated the need for Maori to be involved in the decision-making processes that relate to the health of Maori. This applies in all areas of Maori health. Unless there is an absolute commitment to this premise, there is unlikely to be any improvement in the status of Maori health in the foreseeable future.

List of Contributors

Dr George Barker
George Barker is Manager of the Strategic Analysis Unit, Social Policy and Government Services at The Treasury. Over the past year he has been seconded to the Institute of Policy Studies at Victoria University. He has recently published *Income Distribution in New Zealand.*

Dr Alison Blaiklock
Alison Blaiklock is a Public Health Medicine Specialist with Health Promotion Auckland, who recently helped organise a conference titled Multiple Effects of Poverty on Children and Young People.

Dr Peter Crampton
Peter Crampton is a Senior Lecturer in Public Health in the Department of Public Health at the Wellington School of Medicine, and a Research Fellow at the Health Services Research Centre. His recent research interests include the development of a new small area index of deprivation, and community controlled primary health care services.

Dr Peter Davis
Peter Davis is a Senior Lecturer in medical sociology at the School of Medicine, University of Auckland. His recent research has included collaborative work with others in developing an updated index of occupational class.

Associate Professor Brian Easton
Brian Easton is an economist and social commentator with honorary appointments at the Department of Public Health, Wellington School of Medicine and the Central Institute of Technology. His ongoing interests include macroeconomic policy and social policy.

Dr Philippa Howden-Chapman
Philippa Howden-Chapman is a Senior Lecturer in Health and Social Science in the Department of Public Health at the Wellington School of Medicine. Her recent research interests include collaborative work with

Peter Davis and others in developing an updated index of occupational class, and analysing the effects of the health reforms.

Dr Ichiro Kawachi
Ichiro Kawachi is Assistant Professor of Public Health at Harvard University. He gained his PhD in public health at the Wellington School of Medicine. His recent research interests include examination of the relationships between income distribution, social capital and health in the United States.

Mr Bruce P Kennedy
Bruce Kennedy is an instructor at the Department of Health, Policy and Management at the Harvard School of Public Helath, Boston.

Ms Cynthia Kiro
Cynthia Kiro of Nga Puhi descent is a Lecturer in Social Policy and Social Work at Massey University in Auckland. Her research interests include Maori health policy and practice.

Mr Keith McLeod
Keith McLeod is a mathematical statistician at Statistics New Zealand. He has recently been involved in collaborative work with Peter Davis and Philippa Howden-Chapman and others in the development of an updated index of occupational class.

Mrs Clare Salmond
Clare Salmond is a Senior Lecturer in Biostatistics in the Department of Public Health at the Wellington School of Medicine. Her recent research interests include the development of a new small area index of deprivation.

Professor Peter Saunders
Peter Saunders is Director of the Social Policy Research Centre at the University of New South Wales. He has extensive research experience in the field of social policy, including collaborative work with New Zealand researchers in studies of income distribution.

Dr Robert Stephens
Robert Stephens is a Senior Lecturer with the Public Policy Group at

Victoria University. His recent research interests have included the development of New Zealand poverty lines.

Ms Susan St John

Susan St John is a Senior Lecturer in Economics in the Department of Economics at Auckland University. Her research interests include applied macroeconomics, family policy and superannuation.

Dr Frances Sutton

Frances Sutton has been undertaking health sector quantitative and economic analyses since 1978. She specialises in population-based funding formulae, health sector expenditure analysis and forecasting, and policy costing. She has collaborated in the development of small area indices of deprivation.

Mr Charles Waldegrave

Charles Waldegrave leads the Social Policy Research Unit at the Family Centre, Lower Hutt. He is a joint leader of the New Zealand Poverty Measurement Project and is extensively involved in social policy research analysis and advocacy over a wide range of topics in New Zealand society.

Professor Alistair Woodward

Alistair Woodward is Professor of Public Health at the Wellington School of Medicine. His interests lie in environmental health, epidemiology and public health, and the general question of what makes populations susceptible to illhealth.

List of Conference Participants

Ms Frances Acey, Wellington City Council
Ms Karen Adams, NZ Council of Christian Social Services
Ms Bridget Allan, Hokianga Health Enterprise Trust
Ms Diane Anderson, MAF Policy
Dr John Angus, Social Policy Agency
Mr Derek Asher
Mr Clive Aspin, Department of Public Health, Wellington School of
    Medicine
Mr Brendon Baker, Maori Health Group, Ministry of Health
Dr Michael Baker, ESR
Mr Justin Bangma, Social Policy Agency
Dr George Barker, The Treasury
Ms Juliette Begg, ESR
Ms Heather Bell, Department of Public Health, Wellington School of
    Medicine
Mrs Maraea Bellamy, Maori Health Group, Ministry of Health
Dr Helen Bichan, Australasian Faculty of Public Health Medicine
Dr Alison Blaiklock, Health Promotion Auckland
Dr Tony Blakely, Public Health Medicine Registrar
Ms Sharron Bowers, Department of Public Health, Wellington School of
    Medicine
Mrs Annette Bridgman, The Treasury
Mr Paul Calcott, Economics Department, Victoria University
A/Prof Sally Casswell, Alcohol and Public Health Research Unit
Mr Jit Cheung, Midland Health
Mr Mark Clements, Ministry of Health
Dr Sunny Collings, Wellington School of Medicine
Ms Philippa Conway, National Health Committee
Prof Michael Cooper, Central Institute of Technology
Dr Peter Crampton, Health Services Research Centre
Dr Judith Davey, Victoria University
Dr Nigel Dickson, Southern Regional Health Authority
Ms Roz Dibley, Occupational Safety and Health
Ms Sylvia Dixon, Department of Labour
Ms Trish Donnelly, Central RHA

179

Dr Wendy Drewery, University of Waikato
Mr Andrew Duncan, Ministry of Health
Ms Mavis Duncanson, Health Services Research Centre
Ms Sonja Easterbrook-Smith, Star Bay Associates
A/Prof Brian Easton, Wellington School of Medicine
Ms Nicolette Edgar, Mental Health Commission
Ms Debbie Edwards, North Health
Ms Judy Edwards, Ministry of Health
Ms Caroline Everard, National Maori SIDS Prevention, Department of
    Maori and Pacific Health
Ms Maureen Gillon, RNZCGP
Mrs Judy Glackin, Ministry of Health
Mr Peter Glensor, Health Care Aotearoa
Ms Kawshaliya Gooneratne, Department of Public Health (Masters
    Student)
Ms Michele Grigg, Ministry of Health
Mr Kenneth Hand, The Treasury
Mr Graham Harrison, Harrison Consulting
Ms Ruth Harrison, Clear Ideas
Ms Julie Helean, North Health
Mrs Pauline Hill, Ministry of Health
Dr Andrew Holmes, National Health Committee
Mr Michael Howard, Ministry of Health
Dr Philippa Howden-Chapman, Department of Public Health,
    Wellington School of Medicine
Mr Paora Howe, Te Puni Kokiri
Ms Frances Hughes, Ministry of Health
Dr Tarek Mahmud Hussain, University of Waikato
Prof Barbara Israel, Department of Public Health, Wellington School of
    Medicine
Mr Stephen Jacobs, Wesley Wellington Mission
Dr Ichiro Kawachi, Harvard University
Mr David Kay, Ministry of Health
Mr Julian King, Ministry of Health
Mr Te Kahi Kingi, Te Pumanawa Hauora
Ms Cynthia Kiro, Social Policy and Social Work, Massey University Albany
Jesse Kokaua, Southern Regional Health Authority
Mr Todd Krieble, Ministry of Health (Sector Strategy)

Dr Anthony Kriechbaum, Public Health Service, MidCentral Health
Mr Colin Lewis, Ministry of Maori Development
Mrs Jean Maines, Health Waikato
Dr Murray Malcolm, Consultant
Mrs Cynthia Maling, Central RHA
Mr David Marra, 198 Youth Health Centre
Dr Don Matheson, Tairawhiti Healthcare
Ms Mary McCulloch, Central Regional Health Authority
Dr Fran McGrath, Central Regional Health Authority
Mr Keith McLeod, Statistics New Zealand
Ms Anne McNicholas, ESR
Ms Lesley Middleton, Ministry of Health (Sector Strategy)
Ms Rowena Millmow, Ministry of Health
Mr David Moore, Pharmac
Ms Vivienne Morrell, Ministry of Justice
Ms Jenni Norris, Public Health Association
Dr Pauline Norris, Health Services Research Centre
Dr Fiona North, Department of Preventive and Social Medicine,
    University of Otago
Ms Valerie Norton, ALAC
Dr Michael O'Brien, Department of Social Policy and Social Work,
    Massey University
Mr Des O'Dea
Mr Patrick Ongley, Statistics New Zealand
Ms Justine O'Reilly, National Health Committee
Dr Charles Pain, Midland Regional Health Authority
Dr Julie Park, Anthropology Department, University of Auckland
Ms Lesley Patterson, School of Nursing and Environmental Sciences,
    Wellington Polytechnic
Ms Judy Paulin, Ministry of Justice
Dr David Peacock, Registrar in Public Health Medicine
Dr Julia Peters, AFPHM
Dr Nancy Pollock, Victoria University
Dr Raymond Pong, Northern Health Human Resources Research Unit,
    Laurentian University, Canada
Dr Ian Prior, Department of Public Health, Wellington School of
    Medicine
Gordon Purdie Department of Public Health, Wellington School of
    Medicine

Mr Robin Ransom, AGB McNair
Mihi Ratima, Te Pumanawa Hauora
Dr Judith Reinken, FFMS Consultant
Ms Stephanie Roberts, Clear Ideas
Mrs Marilyn Ross, Health Waikato
Mrs Clare Salmond, Department of Public Health, Wellington School
  of Medicine
Prof George Salmond, Health Services Research Centre
Mr Stephen Salzano, Ministry of Health
Prof Peter Saunders, Social Policy Research Centre
Dr Kay Saville-Smith, CRESA
Dr Bruce A Scoggins, Health Research Council of NZ
Mr David Scott, Central RHA
Ms Jenny Shieff, National Health Committee
Dr Louise Signal, Ministry of Health
Ms Rosemary Simpson, Ministry of Women's Affairs
Mr Neil Smiler, Ministry of Maori Development (Te Puni Kokiri)
Mr Don Smith, National Health Committee
Ms Suzanne Snively, Coopers & Lybrand
Ms Paula Snowden, Ministry of Women's Affairs
Mrs Debbie Sorensen, Ministry of Pacific Island Affairs
Dr Ian St George, Postgraduate Department, Wellington School of
  Medicine
Ms Susan St John, Department of Economics, Auckland University
Dr Bob Stephens, Public Policy Group, Victoria University
Mr Shane Stuart, The Family Centre
Ms Frances Sutton
Dr Boyd Swinburn, National Heart Foundation
Ms Jenni Tarrant, North Health
Ms Julia Tinga, Ministry of Health
Mr Martin Tobias, Ministry of Health
Ms Sarah Turner, Ministry of Health
Mr Charles Waldegrave, Social Policy Research Unit, Family Centre
Mr John Waldon, Te Pumanawa Hauora
Ms Rahira Walsh, Maori Health Group, Ministry of Health
Ms Ingrid Ward, Ministry of Health
Ms Sylvia Watson, Central Regional Health Authority
Dr Gary Whitlock, Clinical Trials Reseach Unit

Ms Brigid Wilson, Southern Regional Health Authority
Ms Wendy Woodhouse, Public Health, Wairarapa Health
Prof Alistair Woodward, Department of Public Health, Wellington
    School of Medicine
Mrs Dawn Young, Waikato Polytechnic Nursing and Health Studies
Mr Ross Young, National Health Committee